WITHDRAWN

Degree Theory
for Equivariant Maps,
the General S^1-Action

Recent Titles in This Series

(See the AMS catalog for earlier titles)

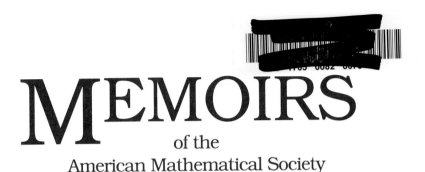

MEMOIRS
of the
American Mathematical Society

Number 481

Degree Theory
for Equivariant Maps,
the General S¹-Action

Jorge Ize
Ivar Massabo
Alfonso Vignoli

November 1992 • Volume 100 • Number 481 (end of volume) • ISSN 0065-9266

American Mathematical Society
Providence, Rhode Island

1991 *Mathematics Subject Classification.*
Primary 58B05; Secondary 34C25, 47H15, 54F45, 55Q91, 58E09.

Library of Congress Cataloging-in-Publication Data

Ize, Jorge, 1946–
  Degree theory for equivariant maps, the general S^1-action/Jorge Ize, Ivar Massabo, Alfonso Vignoli.
    p. cm. – (Memoirs of the American Mathematical Society, ISSN 0065-9266; no. 481)
    Includes bibliographical references.
    ISBN 0-8218-2542-9
    1. Topological degree. 2. Mappings (Mathematics) 3. Homotopy groups. 4. Sphere. I. Massabo, Ivar, 1947– . II. Vignoli, Alfonso, 1940– . III. Title. IV. Series.
QA3.A57 no. 481
[QA612]
510 s–dc20   92-28573
[514′.2]   CIP

Memoirs of the American Mathematical Society

This journal is devoted entirely to research in pure and applied mathematics.

Subscription information. The 1992 subscription begins with Number 459 and consists of six mailings, each containing one or more numbers. Subscription prices for 1992 are $292 list, $234 institutional member. A late charge of 10% of the subscription price will be imposed on orders received from nonmembers after January 1 of the subscription year. Subscribers outside the United States and India must pay a postage surcharge of $25; subscribers in India must pay a postage surcharge of $43. Expedited delivery to destinations in North America $30; elsewhere $82. Each number may be ordered separately; *please specify number* when ordering an individual number. For prices and titles of recently released numbers, see the New Publications sections of the *Notices of the American Mathematical Society*.

Back number information. For back issues see the *AMS Catalogue of Publications*.

Subscriptions and orders should be addressed to the American Mathematical Society, P. O. Box 1571, Annex Station, Providence, RI 02901-1571. *All orders must be accompanied by payment.* Other correspondence should be addressed to Box 6248, Providence, RI 02940-6248.

Memoirs of the American Mathematical Society is published bimonthly (each volume consisting usually of more than one number) by the American Mathematical Society at 201 Charles Street, Providence, RI 02904-2213. Second-class postage paid at Providence, Rhode Island. Postmaster: Send address changes to Memoirs, American Mathematical Society, P. O. Box 6248, Providence, RI 02940-6248.

TABLE OF CONTENTS

ABSTRACT

In the first paper of this series we constructed a general degree theory for maps which are equivariant with respect to a linear action of an arbitrary compact Lie group. This degree takes values in certain equivariant homotopy groups of spheres. These groups were computed for the particular case when the circle group S^1 acts almost semi–freely. In the present paper we consider general S^1–actions, which may differ on the domain and on the range, with isotropy subspaces with one dimension more on the domain. In the special case of self-maps the S^1-degree is given by the usual degree of the invariant part, while for one parameter S^1-maps one has an integer for each isotropy subgroup different from S^1. In particular we recover all the S^1–degrees introduced in special cases by other authors and we are also able to interpret period doubling results on the basis of our S^1–degree. The applications concern essentially periodic solutions of ordinary differential equations.

Key words and phrases .

Equivariant topological degree, S^1–homotopy group of spheres.

ACKNOWLEDGMENTS

Part of the research for this paper was done during visits of the first author to the universities of Cosenza and Roma II.

The support of these universities, of the Consiglio Nazionale delle Ricerche and of the Centro Matematico Vito Volterra, of the University of Roma II, was essential for these visits. The research for the problems with first integrals (sections 5.7 and 6.4) was conducted during a visit of the first author to the University of New England, Australia. Long conversations with E.N. Dancer were quite stimulating. Special thanks are due to Alma Rosa Rodríguez who has beautifully typed and processed the manuscript.

INTRODUCTION

The classical degree theories have been very helpful in the study of nonlinear problems. Many of these present natural symmetries which have been used to simplify, in certain generic cases, the analysis of the structure of these problems. In [**I.M.V**] we have defined a degree theory for equivariant maps in the following context.

Let E and F be two Banach spaces and Γ be a compact Lie group acting linearly, via isometries, on both of them. Let Ω be a bounded open invariant subset of E and f be an equivariant map defined in $\bar{\Omega}$, that is:

$$f(\gamma x) = \tilde{\gamma} f(x)$$

for all x in $\bar{\Omega}$ and γ in Γ ($\tilde{\gamma}$ represents the action on F).

We have studied the equivariant homotopy groups for such mappings and defined the equivariant degree for f relative to Ω, $\deg_\Gamma(f; \Omega)$, to be an element of such groups. All maps and homotopies must be equivariant and all sets must be invariant. The mappings have a certain amount of compactness. This degree has all the properties of a classical degree theory and it reduces to the Leray-Schauder degree when $\Gamma = \{e\}$ the identity, i.e., there is no action.

Furthermore, we have computed this degree when Γ is S^1 and there is just one non-trivial isotropy subgroup (an almost semi-free action), recovering Fuller's degree in the case of autonomous differential equations.

Received by the Editors January 10, 1990 and in revised form September 16, 1991

1

This paper is essentially the continuation of the program announced in [**I.M.V**]. To be more precise we shall dwell here into the computation of the S^1-degree in the case when the actions of S^1 are not necessarily *almost semi-free*.

We shall not assume that the reader is completely familiar with the material expounded in [**I.M.V**]. We shall adopt the same notations of our former work all the way through. We feel, however, that subsuming those items from [**I.M.V**] which are directly related to the subject matter of the present work may be helpfull.

Let us recall first how we defined the Γ-degree in the case when Γ is an arbitrary compact Lie group acting between finite dimensional Euclidean Γ-spaces of possibly different dimensions.

Let Γ act linearly and via isometries on both $\mathbb{R}^m$ and $\mathbb{R}^n$ and let Ω be a Γ-invariant, bounded, open subset of $\mathbb{R}^m$. Assume that $f : \overline{\Omega} \longrightarrow \mathbb{R}^n$ is a Γ-equivariant continuous map such that $f(x) \neq 0$ on $\partial\Omega$. The Γ-degree for f is constructed as follows.

Take a sufficently large closed ball B centered at the origin containing Ω and let $\hat{f} : B \longrightarrow \mathbb{R}^n$ be a Γ-equivariant continuous extension of f. Let N be a bounded open invariant neighborhood of $\partial\Omega$ such that $\hat{f}(x) \neq 0$ for any $x \in \overline{N}$. Define $V = (\Omega \cup N)^c$ and let $\phi : B \longrightarrow [0,1]$ be a Urysohn function such that $\phi(x) = 1$ if $x \in V$ and $\phi(x) = 0$ if $x \in \overline{\Omega}$. Let $F : [0,1] \times B \longrightarrow \mathbb{R} \times \mathbb{R}^n$ be the map defined by

$$F(t,x) = (2t + 2\phi(x) - 1, \hat{f}(x)),$$

where Γ acts trivially on both $[0,1]$ and $\mathbb{R}$.

It is easily seen that $F(t,x) = 0$ only if $x \in \Omega$, $f(x) = 0$ and $t = \frac{1}{2}$. Thus F can be regarded (via standard identification) as a Γ-equivariant map from S^m into S^n. The Γ-degree of f is defined to be the Γ-equivariant homotopy class $[F]_\Gamma$ of F considered as an element of the Γ-equivariant homotopy group of spheres $\Pi_m^\Gamma(S^n)$.

It is not too hard to show that $[F]_\Gamma$ does not depend on its construction and hence our definition of the Γ-degree of f, denoted $deg_\Gamma(f; \Omega)$, is well posed. Moreover, this Γ-degree satisfies all the fundamental properties of a classical topological degree theory:

a)**Existence Property.** If $\deg_\Gamma(f;\Omega)$ is nontrivial, then there exists $x \in \Omega$ such that $f(x) = 0$.

b)**Γ-homotopy invariance property.** Let $f_\tau : \bar\Omega \to \mathbb{R}^n, 0 \le \tau \le 1$, be a continuous one-parameter family of Γ-equivariant maps not vanishing on $\partial\Omega$ for all $\tau \in [0,1]$. Then the Γ-degree, $\deg_\Gamma(f_\tau;\Omega)$, does not depend on $\tau \in [0,1]$.

c)**Excision property.** Let f: $\bar\Omega \to \mathbb{R}^n$ be a continuous Γ-equivariant map such that $f(x) \ne 0$ in

$\Omega\backslash\Omega_0$, where $\Omega_0 \subset \Omega$ is open and Γ-invariant. Then

$$\deg_\Gamma(f;\Omega) = \deg_\Gamma(f_{|\bar\Omega_0};\Omega_0)$$

d)**Suspension property.** Let $f : B_R \to \mathbb{R}^n$ be a Γ-equivariant continuous map not vanishing on ∂B_R. Then

$$\deg_\Gamma(f;B_R) = \Sigma_0([f]_\Gamma)$$

where Σ_0 is the suspension operation.

Moreover, if Σ_0 is a monomorphism and $\deg_\Gamma(f;B_R) = 0$ then f has a non-vanishing Γ-extension to B_R. This is the well-known **Hopf property.**

e)**Additivity property** (up to one suspension). Let $f : \bar\Omega \to \mathbb{R}^n$ be a Γ-equivariant continuous map such that $f(x) \ne 0$ on $\partial\Omega$ and let $\Omega = \Omega_1 \cup \Omega_2$, where Ω_1,Ω_2 are open Γ-invariant subsets of Ω such that $\bar\Omega_1 \cap \bar\Omega_2 = \emptyset$. Then

$$\Sigma_0([F]_\Gamma) = \Sigma_0([F_1]_\Gamma) + \Sigma_0([F_2]_\Gamma),$$

where $[F]_\Gamma, [F_1]_\Gamma$ and $[F_2]_\Gamma$ are the Γ-equivariant homotopy classes induced by $f, f_{|\bar\Omega_1}$, and $f_{|\bar\Omega_2}$ respectively.

In the context of actions considered in the present paper, we shall see that Σ_0 is a monomorphism (cfr. Theorem 3.3). In general this might not be true: the Appendix contains an example where the additivity is true only after a suspension.

f) **Recovering the Brouwer degree.** In the case when $\Gamma = \{e\}$ (i.e., Γ is the trivial group) and $m = n$ the Γ-degree reduces to the classical Brouwer topological degree (here denoted by $\deg_B$). Indeed,

$$\deg_\Gamma(f;\Omega) = [F]_\Gamma = \deg_B(F;(0,1) \times B_R;0) = \deg_B(F;(0,1) \times \Omega;0)$$
$$= \deg_B((2t-1,f);(0,1) \times \Omega;0).$$

Using the product formula of the Brouwer degree we obtain

$$\deg_B((2t-1,f);(0,1) \times \Omega;0) = \deg_B(2t-1;(0,1);0)\deg_B(f;\Omega;0)$$
$$= \deg_B(f;\Omega;0).$$

g) **Universality property.** Let $\Delta_\Gamma(f;\Omega)$ be any other Γ-degree with the properties (a), (b), (c) and such that Σ_0 is a monomorphism. Then, whenever $\Delta_\Gamma(f;\Omega)$ is non trivial, this is also true for $\deg_\Gamma(f;\Omega)$.

In fact, one may compute

$$\Delta_\Gamma(F; I \times B) = \Delta_\Gamma((2t-1,f); I \times \Omega) = \Sigma_0\Delta_\Gamma(f;\Omega),$$

where the first equality is the excision of $I \times B\backslash I \times \Omega$ and the second is the suspension operation. Thus if $\deg_\Gamma(f;\Omega)$ is zero, then F has a non-zero extension from $\partial(I \times B)$ to $I \times B$. From (a), $\Delta_\Gamma(F, I \times B)$ must be trivial and, since Σ_0 is one to one, $\Delta_\Gamma(f;\Omega)$ must be trivial. The converse is not necessarily true unless Δ_Γ satisfies the Hopf property.

h) **Extension to infinite dimension.** The Γ-degree has been extended to the infinite dimensional context and it has been computed, following the same type of ideas used when passing from the Brouwer to the Leray-Schauder degree, by suitable Γ-equivariant finite dimensional approximations. We have considered the following framework.

Let E be a Γ-Banach space and let Ω be a Γ-invariant open bounded subset of $\mathbb{R}^M \times E$. Let

$$f(x,y) = (f_N(x,y), f_\infty(x,y)) : \qquad \Omega \to \mathbb{R}^N \times E$$

be a compact Γ-map such that the Γ-map

$$\Phi(x,y) = (-f_N(x,y), y - f_\infty(x,y))$$

is non-zero on $\partial\Omega$.

Then the Γ-degree of Φ with respect to Ω is an element of the stable group of Γ-homotopies: $\Pi^\Gamma_{M,N}$.

From our definition of $deg_\Gamma(f;\Omega)$ it becomes apparent that everything boils down to computing $\Pi^\Gamma_m(S^n)$. Of course very little is known about these groups for an arbitrary compact Lie group Γ. As a matter of fact the bulk of [**I.M.V**] consists of computing the groups $\Pi^{S^1}_m(S^n)$ when the circle group S^1 acts almost semi-freely (a quick account of these results is given below in the Preliminaries).

As far as the structure of this work is concerned we have the following. Chapter 1 is a concise exposition of some of the results contained in [**I.M.V**] representing the starting point for this paper.

Chapter 2 is devoted to the elementary concept of extension of equivariant maps. This is the essential tool in defining the extension degree. In fact, any linear action of S^1 on $V \equiv \mathbb{R}^{k+1} \times \mathbb{C}^m$ and on $W \equiv \mathbb{R}^{l+1} \times \mathbb{C}^n$ will be of the form:

$$e^{im\varphi}(x, z_1, \ldots, z_m) = (x, e^{im_1\varphi}z_1, \ldots, e^{im_m\varphi}z_m)$$
$$e^{in\varphi}(y, \xi_1, \ldots, \xi_n) = (y, e^{in_1\varphi}\xi_1, \ldots, e^{in_n\varphi}\xi_n),$$

where $m_i, i = 1, \ldots, m$ and $n_j, j = 1, \ldots, n$ can always be assumed to be positive integers. Thus, a S^1–map f from V into W must respect these actions. If f is non–zero on the boundary of an invariant ball B in V, then $[f]_{S^1}$ will be trivial if and only if f has a non–zero S^1-extension to the whole ball B. Now, if H is an isotropy subgroup of S^1 (i.e., H fixes some point of V) and V^H is the fixed point subspace for H, then $f^H \equiv f|V^H$ has its range in W^H, as explained later on.

Suppose that $f^K: S^K \rightarrow W^K\setminus\{0\}$ has a non–zero extension to B^K for all isotropy subgroups K, with $H < K$. Then, by a step by step construction for an explicit S^1–cell decomposition, we have shown (cfr. Theorem 2.1, Corollary 2.1 and Lemma 2.2 below):

Theorem 1.

(i) *If* $\dim_{\mathbb{R}} V^H \leq \dim_{\mathbb{R}} W^H$ *then* $f^H : S^H \to W^H \backslash \{0\}$ *has also a non–zero* S^1*–extension to* B^H. *In particular, if this inequality is true for all subgroups* H, *then the* S^1*–extendability of* f *from* S *to* B *depends only on the invariant part.*

(ii) *If* $\dim_{\mathbb{R}} V^H = \dim_{\mathbb{R}} W^H + \delta$, $\delta = 1$ *or* 2, *then the extension degree of* f^H *is in* $\mathbb{Z}$ *if* $\delta = 1$ *and in* $\mathbb{Z}_2$ *if* $\delta = 2$, *which is the obstruction for the extendability to* B^H. *For* $\delta = 1$ *this degree is well defined (i.e., does not depend on the previous extensions) and a morphism (if* $k = l - 1$ *this is true modulo* $|H| \Pi n_j / \Pi m_j$ *where* m_j *and* n_j *are the modes present in* V^H *and* W^H *respectively and* $|H|$ *is the order of the group* H).

For the reader's convenience and in order to give the flavor of the tools used in the paper, we shall interpret Theorem 1 in a simple context.

Let $k = l + 1, m_1 = m_2 = n_1 = n_2 = 2$ and $m_3 = m_4 = n_3 = n_4 = 3$. The elements of $V = \mathbb{R}^{l+2} \times \mathbb{C}^4$ will have the form (x, Z_2, Z_3) with $x \in \mathbb{R}^{l+2}$, $Z_2 = (z_1, z_2)$ and $Z_3 = (z_3, z_4)$, where $z_j \in \mathbb{C}$ for $j = 1, \ldots, 4$.

In this case the isotropy subgroups of S^1 are:

$H_0 = S^1$ with isotropy subspace $V^{S^1} = \{(x, 0, 0)\}$,

$H_3 = \mathbb{Z}_3$ with isotropy subspace $V^{H_3} = \{(x, 0, Z_3)\}$, corresponding to $\varphi = 2k\pi/3$,

$H_2 = \mathbb{Z}_2$ with isotropy subspace $V^{H_2} = \{(x, Z_2, 0)\}$, corresponding to $\varphi = k\pi$,

$H_1 = \{e\}$ with isotropy subspace $V^{H_1} = V$ and the points with isotropy subgroup H_1 are those with $Z_2 \neq 0$ and $Z_3 \neq 0$.

Note that $\dim_{\mathbb{R}} V^H = \dim_{\mathbb{R}} W^H + 1$ for all isotropy groups H.

Let B be an invariant ball in V and $f : B \to W = \mathbb{R}^{l+1} \times \mathbb{C}^4$ be a S^1-map which is non-zero on the boundary S of B.

For simplicity, the map f will be written as (f_0, f_2, f_3) on the modes of W, that is

$$f_j(x, e^{2i\varphi} Z_2, e^{3i\varphi} Z_3) = e^{ji\varphi} f_j(x, Z_2, Z_3)$$

for $j = 0, 2, 3$.

In particular $f_2(x, 0, Z_3) = 0$ and $f_3(x, Z_2, 0) = 0$, since f^H sends V^H into W^H.

The constrains on the extendability of f from S to all of B, as described in Theorem 1, are contained in the following step by step construction. In order to avoid heavy notations, we shall keep the symbols of the original maps for their extensions.

1st. step. Consider $f^{H_0} : \partial(B^{H_0}) \simeq S^{l+1} \to \mathbb{R}^{l+1}\backslash\{0\}$. Then $f^{H_0}(x, 0, 0) = f_0(x, 0, 0)$ extends to a non-zero map on $B_0 = B^{H_0}$ if and only if its class in $\Pi_{l+1}(S^l)$ is zero.

Note that if $k = l$, then this happens if and only if the Brouwer degree of f^{H_0} is zero.

2nd. step. Assume that $f^{H_0} = f_0$ has a non-zero extension to $B^{H_0} = B_0$. Consider, on the boundary of the relative ball $B_0 \times \{(0, z_4) : 0 \leq z_4 \leq R\}$ in V^{H_3}, the function defined as the extension of f_0 for $z_4 = 0$ and (f_0, f_3) on the rest.

Recall that $f_2(x, 0, Z_3) = 0$. Thus, one has a map from S^{l+2} into $\mathbb{R}^{l+1} \times \mathbb{C}^2\backslash\{0\} \simeq S^{l+4}$ and so an element of $\Pi_{l+2}(S^{l+4}) = 0$. Hence the above map has a non-zero extension to $B_0 \times \{(0, z_4) : 0 \leq z_4 \leq R\}$. One may define an equivariant extension to $B_0 \times \{(0, z_4) : |z_4| \leq R\}$, for $z_4 = e^{i\psi}|z_4|$, through the formula

$$(f_0, f_3)(x, 0, e^{i\psi}|z_4|) = (f_0, e^{i\psi}f_3)(x, 0, |z_4|).$$

Now, on the boundary of $B_0 \times \{(0, z_3, z_4) : 0 \leq z_3 \leq R, |z_4| \leq R\}$, which has dimension $l + 4$, one can consider the non-zero map defined as the previous extension for $z_3 = 0$ and the original map on the rest of the boundary. Clearly, this map defines an element of $\Pi_{l+4}(S^{l+4}) \simeq \mathbb{Z}$. Thus, an obstruction to its extendability is represented by its Brouwer degree.

If this degree is zero, then one has a non-zero extension to

$B_0 \times \{(0, z_3, z_4) : 0 \leq z_3 \leq R, |z_4| \leq R\}$ and, using the action, one obtains a non-zero S^1-map on $B_0 \times \{(0, z_3, z_4) : |z_3| \leq R, |z_4| \leq R\}$ by setting

$$(f_0, f_3)(x, 0, z_3, z_4) = (f_0, e^{i\psi}f_3)(x, 0, |z_3|, e^{-i\psi}z_4)$$

where $z_3 = e^{i\psi}|z_3|$.

Note that if $k = l$ then, in this last step, the obstruction would be in $\Pi_{l+3}(S^{l+4}) = 0$, that is one has always a non-zero equivariant extension to V^{H_3}.

It is clear that the extendability to V^{H_2} can be treated in the same fashion as for V^{H_3}. Once this is done, we are left with

3rd. step. Assume that one has an equivariant extension to $V^{H_3} \cup V^{H_2}$. Its extendability to $V^{H_1} = V$ is treated in the following way. First consider the set

$$D = B_0 \times \{(0, z_2, z_3, z_4) : 0 \le z_2 \le R, 0 \le z_3 \le R, |z_4| \le R\}.$$

The boundary of D is a sphere of dimension $l+5$ and, on it, one has a map with values in $\mathbb{R}^{l+1} \times \mathbb{C}^4 \backslash \{0\} \simeq S^{l+8}$. Thus, this map has an extension to all of D. Observe that z_2 is fixed, under the S^1-action, for $\varphi = k\pi$ and that

$$f(x, Z_2, -Z_3) = (f_0, f_2, -f_3)(x, Z_2, Z_3).$$

Thus, the extension can be viewed, through this $\mathbb{Z}_2$-action, as a map defined on the set

$$B_0 \times \{(0, z_2, z_3, z_4) : 0 \le z_2 \le R, -R \le z_3 \le R, |z_4| \le R\}.$$

Together with the original map one obtains a map which is non-zero on the boundary of the ball $B_0 \times \{(0, z_2, z_3, z_4) : 0 \le z_2 \le R, |z_4| \le R, |z_3| \le R, \text{Arg } z_3 \in [0, \pi]\}$, that is a map from S^{l+6} into S^{l+8}. Hence it has a non-zero extension to the above ball. Furthermore, one may extend the map to the points with $\text{Arg } z_3 \in [\pi, 2\pi]$ through the $\mathbb{Z}_2$-symmetry as well as to the points $z_2 \in \mathbb{C}$ using the formula

$$(f_0, e^{i\psi} f_2, e^{3i\psi/2} f_3)(x, |z_2|, e^{-3i\psi/2} Z_3), \text{ for } z_2 = e^{i\psi}|z_2|.$$

Note that this map is well defined since if one changes ψ to $\psi + 2\pi$ one changes Z_3 into $-Z_3$ and one recovers the finite $\mathbb{Z}_2$-symmetry.

Finally, consider the ball

$$B_0 \times \{(z_1, z_2, z_3, z_4) : 0 \le z_1 \le R, |z_2| \le R, 0 \le z_3 \le R, |z_4| \le R\}.$$

On the boundary of this set, which is a sphere of dimension $l+7$, one has a map valued in S^{l+8}, so that it can be extended to all the ball. Moreover, taking into account the $\mathbb{Z}_2$-symmetry, one obtains on extension for $-R \le z_3 \le R$.

Hence the map is defined on the boundary of the ball

$$D = B_0 \times \{(z_1, z_2, z_3, z_4) : 0 \le z_1 \le R, |z_2| \le R, |z_4| \le R,$$
$$- R \le z_3 \le R, \, \mathrm{Arg}\, z_3 \in [0, \pi]\}.$$

The boundary of D is a sphere of dimension $l + 8$, thus, one has a map from S^{l+8} into itself and an obstruction represented by its Brouwer degree.

If this degree is zero one can complete the extension of the map for $\mathrm{Arg}\, z_3 \in [\pi, 2\pi]$ through the $\mathbb{Z}_2$-symmetry and to $z_1 \in \mathbb{C}$ with the formula

$$(f_0, e^{i\psi} f_2, e^{3i\psi/2} f_3)(x, |z_1|, e^{-i\psi} z_2, e^{-3i\psi/2} z_3)$$

where $z_1 = e^{i\psi} |z_1|$.

Finally, note that if $k = l$, one never gets to the point with the same dimensions and so there is always an equivariant extension.

In Chapter 3 we compute the S^1–homotopy group of spheres for the case $\delta = 1$, δ as in Theorem 1, under the following conditions on the actions: $k = l + 1 - 2p, p \ge 0, n_j = k_j m_j,$ for $j = 1, \ldots, m - p = n, n_j$ are multiples of m_r for all $r = n + 1, \ldots, m$. Then (cfr. Theorems 3.1 and 3.2), by exhibiting generators, we have:

Theorem 2.

$$\Pi^{S^1}_{k+2n}(S^{l+2n}) = \begin{cases} \mathbb{Z}, & \text{if } p > 1 \\ \mathbb{Z}_{(\Pi k j) m_0 / m_m}, & \text{if } p = 1 \\ \Pi_k(S^l) \times \mathbb{Z} \times \ldots \times \mathbb{Z}, & \text{if } p = 0 \end{cases}$$

where m_0 is the largest common divisor of $m_1, \ldots, m_n$ and one has one $\mathbb{Z}$ for each isotropy subgroup H such that $\dim_{\mathbb{R}} V^H = \dim_{\mathbb{R}} W^H + 1$.

In the previous example, one has

$$\Pi^{S^1}_{k+4}(S^{l+4}) = \Pi_{l+1}(S^l) \times \mathbb{Z} \times \mathbb{Z} \times \mathbb{Z}$$

with one $\mathbb{Z}$ corresponding to each $H_j, j = 1, 2, 3$. The elements in each $\mathbb{Z}$ are the Brouwer degrees which are the obstructions for extendability (once one has constructed the extensions for the larger isotropy subgroups).

The remaining part of Chapter 3 is devoted to a complete treatment of the suspension operation and to the relation with the set of all possible degrees of Fuller type as defined in [**I.M.V**].

In Chapter 4 we deal with the degree of S^1–maps both in the finite and infinite dimensional cases. We prove in particular (cfr. Theorem 4.1).

Theorem 3. *Any element of* $\Pi^{S^1}_{k+2m}(S^{l+2n})$ *is achieved as the* S^1*–degree of a map defined on* $\overline{\Omega}$ *provided that the component* d_j *(in the case when* $p=0$*) corresponding to* H_j *is 0 if* $\Omega^{H_j} = \emptyset$ *and* $d_0 = 0$ *if* $l = 2$ *(corresponding to the invariant part). In all cases the* S^1*–degree is fully additive.*

We proceed further by proving that our S^1–degree coincides with the S^1–degree defined in [**D.G.J.M**] in the case considered by these authors, that is if $k = l+1, k_j = 1$.

In Chapter 5, under these conditions for $p = 0$, we compute the index of an isolated non–stationary orbit (of hyperbolic type) by relating the components d_j to the indices of the different Poincaré sections constructed in [**I.M.V**]. In this case at most two of the components d_j are non–zero: d_0 corresponding to H_0 the isotropy subgroup of the orbit and d_1 corresponding to H_1 with $|H_0| = 2|H_1|$ (if $|H_0|$ is even). This result is then applied to the period doubling phenomenon and to examples of autonomous differential equations. One recovers and extends the Fuller degree, the degrees defined by Dancer for gradient maps and by Dancer and Toland for equations with first integrals.

In fact autonomous differential equations (and more generally evolution equations) are a natural source of problems with a S^1-action and with an additional parameter. For example, if one looks for periodic solutions of the system

$$\frac{dX}{d\tau} = f(X) , \ X \in \mathbb{R}^l,$$

which, through the time scaling $t = \nu\tau$, gives the equivalent system

$$\nu\frac{dX}{dt} = f(X),$$

then $2\pi/\nu$-periodic solutions of the first system will correspond to 2π-periodic solutions of the second.

More generally, one may look at 2π-periodic solutions of

$$\frac{dX}{dt} = f(X,\nu),$$

in the space of 2π-periodic functions with a certain regularity.

The action of S^1 is through the time translation: $X(t) \to X(t + \varphi)$.

One may also set the problem in terms of Fourier series, that is, write $X(t) = \Sigma_{-\infty}^{\infty} X_n e^{int}$, with X_n in $\mathbb{C}^l$, X_0 in $\mathbb{R}^l$ and $X_{-n} = \bar{X}_n$. One obtains an equivalent equation:

$$i\nu n\, X_n - f_n(X_0, X_1, X_2, \ldots, \nu) = 0, \quad n = 0, 1, 2, \ldots$$

with $f_n(X_0, X_1, \ldots, \nu) = \frac{1}{2\pi} \int_0^{2\pi} f(X(t), \nu) e^{-int} dt$.

Here the action of S^1 on the Fourier series takes the form

$$e^{i\varphi}(X_0, X_1, X_2, \ldots) \equiv (X_0, e^{i\varphi} X_1, e^{2i\varphi} X_2, \ldots)$$

and $f_n(X_0, e^{i\varphi} X_1, e^{2i\varphi} X_2, \ldots) = e^{in\varphi} f_n(X_0, X_1, X_2, \ldots)$, due to the autonomous character.

One also finds S^1-symmetries in elliptic equations on domains which have a spatial S^1-symmetry (here the extra parameter may be the speed of a travelling wave for hyperbolic equations) and in many other problems.

Similarly, in Chapter 6, we compute the index of an isolated orbit of stationary solutions and apply the result to bifurcation of S^1–maps and to general Hopf bifurcation problems for differential equations.

Finally, in Chapter 7, we show how to obtain the orbit index of [M.Y] from the S^1–degree for autonomous differential equations.

CHAPTER ONE

PRELIMINARIES

In this part of the paper we shall recall some basic tools of [**I.M.V**].

1.1. S^1-actions.

Let $\mathbb{R}^M$, $\mathbb{R}^N$ be linear representations of S^1 with fixed point subspaces $\mathbb{R}^k$ and $\mathbb{R}^l$ respectively. Then the S^1-action on the orthogonal complements of those subspaces gives them a complex structure. Thus $\mathbb{R}^M = \mathbb{R}^k \times \mathbb{C}^m$, $\mathbb{R}^N = \mathbb{R}^l \times \mathbb{C}^n$ and the action is represented by

$$
(1.1) \qquad
\begin{aligned}
\mathbf{e}^{im\varphi} z &:= \mathbf{e}^{im\varphi}(z_1, \cdots, z_m) = (e^{im_1\varphi} z_1, \cdots, e^{im_m\varphi} z_m) \\
\mathbf{e}^{in\varphi} \xi &:= \mathbf{e}^{in\varphi}(\xi_1, \cdots, \xi_n) = (e^{in_1\varphi}\xi_1, \cdots, e^{in_n\varphi}\xi_n),
\end{aligned}
$$

where z_i, ξ_j are complex numbers and m_i, n_j are non-zero integers, $i = 1, \cdots, m$ and $j = 1, \cdots, n$. By taking conjugates if necessary, we assume without loss of generality, that m_i and n_j are positive. We set $M' = \prod_{i=1}^{m} m_i$, $N' = \prod_{i=1}^{n} n_i$ and let m_0 be a common divisor of m_i, $i = 1, \ldots, m$.

The elements of $\mathbb{R}^M$ are written as $(x_0, z) \in \mathbb{R}^k \times \mathbb{C}^m$ and any S^1-map from an invariant subset of $\mathbb{R}^M = \mathbb{R}^k \times \mathbb{C}^m$ into $\mathbb{R}^N = \mathbb{R}^l \times \mathbb{C}^n$ has the form $(\Phi_0(x_0, z), \Phi(x_0, z))$ where $\Phi_0(x_0, \mathbf{e}^{im\varphi} z) = \Phi_0(x_0, z)$ and $\Phi(x_0, \mathbf{e}^{im\varphi} z) = \mathbf{e}^{in\varphi}\Phi(x_0, z)$. Note that due to the equivariance of Φ, the integer m_0 divides n_j, $j = 1, \cdots, n$, provided that the $j - th$ component of Φ is not identically equal to zero.

1.2. Almost semi-free actions.

Let $\Pi_{k+2m}^{S^1}(S^{l+2n})$ be the S^1-homotopy group of equivariant maps from $\partial([0,1] \times D)$ into $\mathbb{R} \times \mathbb{R}^N \setminus \{0\}$ of the form $(f_0(x,z), \Phi_0(x,z), \Phi(x,z))$ where D is the set $D = \{(x_0, z) : \|x_0\| \leq R_0, \|z\| \leq R\}$ and the variable x stands for the pair $(t, x_0) \in [0,1] \times \mathbb{R}^k$. For the definition and elementary properties of these homotopy groups see [**I.M.V**, Appendix A].

Let $\mathbb{C}_\bullet^m$ denote $\mathbb{C}^m$ with the S^1-action described by

$$e^{im_0\varphi} Z := e^{im_0\varphi}(Z_1, \cdots, Z_m) = (e^{im_0\varphi} Z_1, \cdots, e^{im_0\varphi} Z_m),$$

which reduces to the standard free-action of S^1 when $m_0 = 1$.

Then, leaving S^1 to act trivially on $\mathbb{R}^k$, the S^1-action on $\mathbb{R}^k \times \mathbb{C}_\bullet^m$ is called **almost semi-free**.

Now, let $\Theta : \mathbb{R}^k \times \mathbb{C}_\bullet^m \to \mathbb{R}^k \times \mathbb{C}^m$ be the S^1-map defined by

$$(1.2) \qquad \Theta(x_0, Z_1, \cdots, Z_m) = (x_0, Z_1^{\mu_1}, \cdots, Z_m^{\mu_m}), \quad \text{where} \quad m_0\mu_i = m_i,$$

and let $\Pi_{k+2m,\bullet}^{S^1}(S^{l+2n})$ be the group of all equivariant homotopy classes of S^1-maps from the boundary of the set $D' = \{(t, x_0, Z) : t \in [0,1], \|x_0\| \leq R_0, \|Z\| \leq R\}$ into $\mathbb{R} \times \mathbb{R}^l \times \mathbb{C}^n \setminus \{0\}$.

The map Θ induces a group homomorphism, see [**I.M.V**, Lemma 4.1]:

$$\Theta^* : \Pi_{k+2m}^{S^1}(S^{l+2n}) \longrightarrow \Pi_{k+2m,\bullet}^{S^1}(S^{l+2n}).$$

Another useful result that facilitates the computation of the degree of S^1-maps is the following. Let Ω be an open and bounded S^1-invariant subset of $\mathbb{R}^k \times \mathbb{C}^m$ and $f : \partial\Omega \to \mathbb{R}^l \times \mathbb{C}^n \setminus \{0\}$ be S^1-equivariant (the action is the one described by (1.1)). Then

$$(1.3) \qquad \Theta^* \deg_{S^1}(f; \Omega) = \deg_{S^1}(f \circ \Theta; \Theta^{-1}(\Omega))$$

(cfr. [**I.M.V**, Lemma 4.2]).

1.3 Equivariant homotopy.

The following morphism of groups relates the equivariant homotopy class of a map with its invariant part. More precisely, let $F(t, x_0, z) = (f_0, \Phi_0, \Phi)(t, x_0, z)$ be a S^1-map from $\partial([0,1] \times D)$ into $\mathbb{R} \times \mathbb{R}^l \times \mathbb{C}^n \setminus \{0\}$ where D is the set defined above. Due to the S^1-action, we have that $\Phi(t, x_0, 0) = 0$. Thus, to the S^1-map F we can associate the S^1-*invariant* map $(f_0, \Phi_0) : \partial([0,1] \times \{x_0 \in \mathbb{R}^k : \|x_0\| \leq R_0\}) \longrightarrow \mathbb{R} \times \mathbb{R}^l$ defined as $(f_0, \Phi_0)(t, x_0) = (f_0, \Phi_0)(t, x_0, 0)$. This assignment is denoted by P. The map P induces a group homomorphism

$$P_* : \Pi^{S^1}_{k+2m,\bullet}(S^{l+2n}) \longrightarrow \Pi_k(S^l)$$

which is onto if $m \leq n$. Furthermore, assuming $m \leq n$ and letting $F = (f_0, \Phi_0, \Phi)$ to represent an element of $\Pi^{S^1}_{k+2m,\bullet}(S^{l+2n})$, the assignment

$$[F]_{S^1} \mapsto (P_*([F]_{S^1}), [F]_{S^1} - [\tilde{F}]_{S^1})$$

from $\Pi^{S^1}_{k+2m,\bullet}(S^{l+2n})$ onto $\Pi_k(S^l) \times \ker P_*$ is an isomorphism, where $\tilde{F} = (f_0, \Phi_0, \tilde{\Phi})$ with $\tilde{\Phi}(t, x_0, Z) = t(1-t)(Z_1^{\frac{n_1}{m_0}}, \cdots, Z_m^{\frac{n_m}{m_0}}, \underbrace{0, \cdots, 0}_{n-m})$

(see [**I.M.V**, Lemma 4.3]).

In [**I.M.V**] we used, for computational purposes, the following S^1-map

$$\beta : \mathbb{R}^l \times \mathbb{C}^n \longrightarrow \mathbb{R}^l \times \mathbb{C}^n_\diamond$$

defined as

$$\beta(y_0, \xi_1, \ldots, \xi_n) = (y_0, \xi_1^{N'/n_1}, \ldots, \xi_n^{N'/n_n})$$

where $\mathbb{C}^n_\diamond$ stands for $\mathbb{C}^n$ with the S^1-action given by

$$e^{iN'\varphi}(\xi_1, \cdots, \xi_n) := (e^{iN'\varphi}\xi_1, \cdots, e^{iN'\varphi}\xi_n).$$

It is easy to check that the map β induces the group homomorphism

$$\beta_* : \Pi^{S^1}_{k+2m}(S^{l+2n}) \longrightarrow \Pi^{S^1}_{k+2m}(S^{l+2n,\diamond})$$

defined by $\beta_*([F]_{S^1}) = [\beta \circ F]_{S^1}$, where $\Pi^{S^1}_{k+2m}(S^{l+2n,\diamond})$ is the group of all equivariant homotopy classes of S^1-maps with range in $\mathbb{R} \times \mathbb{R}^l \times \mathbb{C}^n_\diamond$.

One of our goals here is to describe the left half of the following commutative diagram.

$$
\begin{array}{ccccccc}
& & & \Theta^* & & & \\
\Pi^{S^1}_{k+2m} & (S^{l+2n}) & \xrightarrow{\quad\quad} & & & \Pi^{S^1}_{k+2m,\bullet} & (S^{l+2n}) \\
\Big| & & P_* \searrow & & \nearrow P_* & \Big| & \\
\beta_* \Big| & & & \Pi_k(S^l) & & \Big| & \beta_* \\
\Big\downarrow & & P_* \nearrow & & \searrow P_* & \Big\downarrow & \\
\Pi^{S^1}_{k+2m} & (S^{l+2n,\diamond}) & \xrightarrow{\quad\quad} & & & \Pi^{S^1}_{k+2m,\bullet} & (S^{l+2n,\diamond}) \\
& & & \Theta^* & & &
\end{array}
$$

In [I.M.V] we have completely determined the right half of the diagram when $m = 1$ or $m \leq n + 1 - \frac{k-l}{2}$. This has been done by studying the kernel of the homomorphism P_* through an equivariant extension procedure that we summarize below.

1.4. The extension degree.

Let $I = [0,1]$, $B_0 = \{x_0 \in \mathbb{R}^k : \|x_0\| \leq R_0\}$ and $B = \{Z \in \mathbb{C}^m_\bullet : \|Z\| \leq R\}$. Until further notice, we will consider S^1-equivariant maps

$$
F : \partial(I \times B_0 \times B) \longrightarrow \mathbb{R} \times \mathbb{R}^l \times \mathbb{C}^n \setminus \{0\}
$$

of the following particular form

$$
F(t, x_0, Z) = (f_0, \Phi_0, \Phi)(t, x_0, Z)
$$

with

$$
F(t, x_0, e^{im_0\varphi}Z) = e^{in\varphi}F(t, x_0, Z) .
$$

Recall that due to the equivariance of F the $n_j's$ are multiples of m_0. Assume that $P_*([F]_{S^1}) = 0$, i.e., the invariant part of F, $(f_0, \Phi_0) : \partial(I \times B_0) \longrightarrow \mathbb{R} \times \mathbb{R}^l \setminus \{0\}$ extends to a nonvanishing map $(\tilde{f}_0, \tilde{\Phi}_0) : I \times B_0 \longrightarrow \mathbb{R} \times \mathbb{R}^l \setminus \{0\}$.

Consider the map $\tilde{F}_1 : \partial(I \times B_0 \times \{r \in \mathbb{R} : 0 \leq r \leq R\}) \longrightarrow \mathbb{R} \times \mathbb{R}^l \times \mathbb{C}^n \setminus \{0\}$ defined by

$$\tilde{F}_1 = \begin{cases} (\tilde{f}_0, \tilde{\Phi}_0, 0) & \text{for } r = 0 \\ F & \text{for } 0 < r \leq R. \end{cases}$$

Clearly, the map $\tilde{F}_1$ gives an element $[\tilde{F}_1]$ in the ordinary homotopy group $\Pi_{k+1}(S^{l+2n})$ and $\tilde{F}_1$ admits a nonvanishing extension to $I \times B_0 \times \{r \in \mathbb{R} : 0 \leq r \leq R\}$ *if and only if* $[\tilde{F}_1] = 0$ (one has always an extension if $k + 1 < l + 2n$).

Suppose now that $(\tilde{f}_0, \tilde{\Phi}_0, \tilde{\Phi})$ is such an extension of $\tilde{F}_1$. Then we may define a S^1-equivariant extension F_1 of $\tilde{F}_1$ to $I \times B_0 \times (B \cap \{Z_2 = \cdots = Z_m = 0\})$ by setting for $Z_1 = e^{i\psi}|Z_1|$

$$F_1(t, x_0, e^{i\psi}|Z_1|) = (\tilde{f}_0, \tilde{\Phi}_0, \cdots, e^{i\frac{n_j}{m_0}\psi}\tilde{\Phi}_j, \cdots)(t, x_0, |Z_1|).$$

Note that $F_1(t, x_0, e^{i(\psi+2k\pi)}|Z_1|) = F_1(t, x_0, e^{i\psi}|Z_1|)$ since n_j is a multiple of m_0 for $j = 1, \cdots, n$. Clearly, F_1 is a nonvanishing extension of F to $I \times B_0 \times (B \cap \{Z_2 = \cdots = Z_m = 0\})$ and it is easy to check that the map F_1 is S^1-equivariant.

Now suppose that the map F has been extended S^1-equivariantly to $I \times B_0 \times (B \cap \{Z_h = Z_{h+1} = \cdots = Z_m = 0\})$ via the S^1-map $F_{h-1}(t, x_0, Z_1, \cdots, Z_{h-1})$. Then on the topological sphere $\partial(I \times B_0 \times (B \cap \{Z_{h+1} = \cdots = Z_m = 0, Z_h \in \mathbb{R}, 0 \leq Z_h \leq R\}))$ one may consider the map $\tilde{F}_h$ defined by

$$\tilde{F}_h = \begin{cases} F_{h-1} & \text{on } I \times B_0 \times (B \cap \{Z_h = Z_{h+1} = \cdots = Z_m = 0\}) \\ F & \text{on the rest of the sphere.} \end{cases}$$

Now the map $\tilde{F}_h$ can be extended to the set $I \times B_0 \times (B \cap \{Z_{h+1} = \cdots = Z_m = 0, Z_h \in \mathbb{R}, 0 \leq Z_h \leq R\})$ if it is homotopically trivial in $\Pi_{k+2h-1}(S^{l+2n})$. If this is the case and $(\tilde{f}_0, \tilde{\Phi}_0, \tilde{\Phi})$ is a nonvanishing extension of $\tilde{F}_h$, one obtains a S^1-equivariant extension F_h of F_{h-1} to the set $I \times B_0 \times (B \cap \{Z_{h+1} = \cdots = Z_m = 0\})$ by setting

$$F_h(t, x_0, Z_1, \cdots, Z_{h-1}, e^{i\psi}|Z_h|) =$$
$$= (\tilde{f}_0, \tilde{\Phi}_0, \cdots, e^{i\frac{n_j}{m_0}\psi}\tilde{\Phi}_j, \cdots)(t, x_0, e^{-i\psi}Z_1, \cdots, e^{-i\psi}Z_{h-1}, |Z_h|)$$

for $Z_h = e^{i\psi}|Z_h|$.

It is easy to see that the map F_h is a S^1-equivariant extension of F to the set $I \times B_0 \times (B \cap \{Z_{h+1} = \cdots = Z_m = 0\})$, using as above the facts that F_{h-1} is equivariant and n_j is a multiple of m_0.

Clearly, the procedure just described may be repeated automatically as long as $k+2h-1 < l+2n$, with $1 \le h \le m$, but it will require a particular consideration as soon as the strict inequality is violated, that is for $h = h_c$, such that $k + 2h_c - 1 = l + 2n$.

Namely, we shall denote by $\deg_K(F)$ the ordinary homotopy class of the map $\tilde{F}_{h_c}$ defined above, where K stands for the kernel of P_*. Therefore,

$$\deg_K(F) \in \begin{cases} \Pi_{k+1}(S^{l+2n}), & \text{if } k-l \ge 2n-1 \ (h_c = 1) \\ \Pi_{l+2n}(S^{l+2n}) \cong \mathbb{Z}, & \text{if } |k-l| \text{ is odd} \\ \Pi_{l+2n+1}(S^{l+2n}) & \text{if } |k-l| \text{ is even .} \end{cases}$$

Note that the degree $\deg_K(F)$ *depends a priori* on the different extensions that one has to perform before arriving at the critical level h_c. That *this is in fact not so* was proved in [**I.M.V**, Theorem 4.6]. But, first of all, we would like to point out some simple consequences of the above construction.

1.5. Equivariant homotopy groups of spheres.

To this task let us introduce the following notations. Given a S^1-equivariant map $F = (f_0, \Phi_0, \Phi)$ let $\chi_*([F])$ denote the class of F in $\Pi_{k+2m}(S^{l+2n})$, i.e., when one forgets the group action. The main results of [**I.M.V**] used in the present paper may be stated as follows.

Theorem 1.1.

(i) If $1 \le m < h_c$, then $P_*: \Pi^{S^1}_{k+2m,\bullet}(S^{l+2n}) \to \Pi_k(S^l)$ is an isomorphism.

(ii) $\deg_K: \mathrm{Ker}\, P_* \to \Pi_{k+2h_c+1}(S^{l+2n})$ is a group homeomorphism (in particular it is independent of the previous extensions) except in the following cases

(a) $k = l-1$, $h_c = n+1$ and $N' = m_0^n$

(b) $k = l-1$, $h_c = n+1$ and $N' > m_0^n$

(c) $k \ge l-1$, $h_c = 1$ and $n = 0$

(d) $|k-l|$ even, $h_c > 1$, $k+l+n > 1$ and $\sum_{j=1}^n \frac{n_j}{m_0} + h_c$ odd

(e) $k = l = 0$ and $n = 1$.

Moreover, there is always an extension to the set $I \times B_0 \times (B \cap \{Z_{h_c+1} = \ldots = Z_m = 0\})$ if either (a) or (c) or (d) holds. If either (b) or (e) holds then $\deg_K$ is unique modulo $\frac{N'}{m_0^n}$.

(iii) If $h_c = 1$, then

$$\Pi^{S^1}_{k+2h_c,\bullet}(S^{l+2n}) \cong \begin{cases} 0, & \text{if } n = 0 \\ \Pi_k(S^l) \times \Pi_{k+1}(S^{l+2n}), & \text{if } n > 0 \end{cases}.$$

(iv) If $h_c > 1, |k - l|$ odd and $n \geq 0$, then

$$\Pi^{S^1}_{k+2h_c,\bullet}(S^{l+2n}) \cong \begin{cases} \Pi_k(S^l) \times \mathbb{Z} & \text{if } k \neq l - 1 \\ \mathbb{Z}_{N'/m_0^n} & \text{if } k = l - 1 \end{cases}$$

Furthermore,

$$\chi_*(\Theta^* f) = \deg_K(\Theta^* f)(\sum_{j=1}^{n} \frac{n_j}{m_0} + m + \deg_K(\Theta^* f))\Sigma^{l+2n}\eta$$

where η is the Hopf map, giving the exact image of the equivariant group under the morphism χ_*.

These results, together with similar assertions for the other cases ($|k - l|$ even and a more complete description of χ_*) are contained in [**I.M.V**, Theorems 4.4, 4.6, Corollaries 4.7, 4.8].

1.6. Equivariant degree in the almost semi-free case.

We shall also need the description of the range of the degree for an almost semi–free S^1–action (see [**I.M.V**, Theorem 4.9]), in the cases considered here. Namely,

Theorem 1.2.

(i) *If $\overline{\Omega}^{S^1} = \emptyset$, then $P_*[(2t + 2\phi - 1, \Phi_0, \Phi)] = 0$ and any element in $\mathrm{Ker}P_*$ is obtained (if $h_c = 1$ any suspension is attained), except in two cases, both with $k = 0$ and $m = 2$: $l = 2$ and $n = 0$ or $l = 0$ and $n = 1$.*

(ii) *If $\overline{\Omega}^{S^1} \neq \emptyset$, then any element of $\Pi^{S^1}_{k+2h_c,\bullet}(S^{l+2n})$ is obtained as the S^1–degree of an equivariant map on $\overline{\Omega}$ (for the elements of $\mathrm{Im}P_*$, any suspension is achieved, while for the elements of $\mathrm{Ker}P_*$ the same exceptions as in (i) hold).*

CHAPTER TWO

EXTENSIONS OF S^1 - MAPS

As in [**I.M.V**], our first task is to explore under which circumstances a map has a non-zero S^1 - extension. This is, in a certain sense, close to elementary obstruction theory since we shall give a cell decomposition, find an obstruction for the extension, define from this obstruction a degree which will characterize the extendability to an isotropy subspace and which will be the keystone for the construction of the general S^1–degree. Our approach is completely constructive and requires only a dimension argument and some considerations from [**I.M.V**].

2.1. The Fundamental Cell Lemma.

The following result, close to *Riemann surface theory*, will be fundamental to describe the extension procedure when we have to consider an action $\mathbb{Z}_p \times \mathbb{Z}_q$ with p, q not coprime.

Lemma 2.1. *Let $p, m_2, \ldots, m_m$ be positive integers with largest common divisor m_0 (l.c.d.)denoted by $(p : m_2 : \ldots : m_m)$. Let $\mathbb{Z}_p$ act on T^{m-1}, the $(m-1)$–dimensional torus, by*

$$\gamma^k(z_2, \ldots, z_m) = (e^{i2\pi k m_2/p} z_2, \ldots, e^{i2\pi k m_m/p} z_m)$$

where $k = 0, \ldots, p-1$ and $z_j = e^{i\varphi_j}$ with $\varphi_j \in [0, 2\pi)$, $j = 2, \ldots, m$.

Let $l_j = (p : m_2 : \ldots m_j)$ be the largest common divisor of $p, m_2, \ldots, m_j$ (thus $l_1 = p$ and $l_m = m_0$) and let $\Delta = \{(\varphi_2, \ldots, \varphi_m) : 0 \le \varphi_j < 2\pi l_j/l_{j-1}, j = 2, \ldots, m\}$. Then $\{\gamma^k \Delta\}$, $k = 0, \ldots, \frac{p}{m_0} - 1$, cover properly T^{m-1} (i.e., in a $1-1$ fashion).

Proof. Note at first that if $\gamma^k z = z$, then km_j/p is an integer n_j, thus $km_j/m_0 = n_j p/m_0$. If p/m_0 has a factor which is not in k then it must be in m_j/m_0 for all j's and then m_0 is not the $l.c.d.$. Thus k must be a multiple of p/m_0. Hence, there is just one isotropy orbit type of the action.

If $\gamma^{k_1} \tilde{z} = \gamma^{k_2} z$, then $\tilde{z} = \gamma^{k_2-k_1} z = \gamma^k z$, that is

$$\varphi_j + 2\pi k m_j/p = \tilde{\varphi}_j \ [mod.2\pi], \ 0 \le \varphi_j, \tilde{\varphi}_j < 2\pi l_j/l_{j-1}$$
$$= \tilde{\varphi}_j + 2\pi p_j.$$

Thus, $|\varphi_2 - \tilde{\varphi}_2| < 2\pi \frac{l_2}{l_1} = 2\pi \frac{l_2}{p}$.

$$|\varphi_2 - \tilde{\varphi}_2| = 2\pi |p_2 \frac{p}{l_2} - k\frac{m_2}{l_2}| \frac{l_2}{p}.$$

Hence, since $(p\, p_2 - k\, m_2)/l_2$ is an integer, then $\varphi_2 = \tilde{\varphi}_2$ and $km_2/l_2 = pp_2/l_2$. Since m_2/l_2 and p/l_2 are coprime, then k must be a multiple of p/l_2. Therefore $k = k_2 p/l_2$.

Arguing by induction, we assume then that $\varphi_{j-1} = \tilde{\varphi}_{j-1}$ and $k = k_{j-1} p/l_{j-1}$. Now,

$$|\varphi_j - \tilde{\varphi}_j| = 2\pi |p_j \frac{l_{j-1}}{l_j} - k_{j-1}\frac{m_j}{l_j}| l_j/l_{j-1} < 2\pi l_j/l_{j-1}$$

(recall that by definition l_j is the largest common factor of m_j and l_{j-1}), thus $\varphi_j = \tilde{\varphi}_j$ and $k_{j-1} m_j/l_j = p_j l_{j-1}/l_j$. Since m_j/l_j and l_{j-1}/l_j are coprime, we have that $k_{j-1} = k_j l_{j-1}/l_j$ and $k = k_j p/l_j$. Hence, for $j = m$, $z = \tilde{z}$ and k is a multiple of p/m_0. Then, we may conclude that the images $\gamma^k \Delta$ of $\Delta (k = 0, \ldots, \frac{p}{m_0} - 1)$ do not intersect each other.

Moreover, $\gamma^k \bar{\Delta}$ is (in $\mathbb{R}^{m-1}$) a translate of $\bar{\Delta}$ and, modulo 2π, consists of several pieces of cubes with same volume as $\bar{\Delta}$, that is $(2\pi)^{m-1} l_m/l_1$. Thus, the total volume of all the cells (for $k = 0, \ldots, \frac{p}{m_0} - 1$) (since we have already proved that the interiors do not intersect) is $(2\pi)^{m-1}$. That is all points are covered by the translates of $\bar{\Delta}$. Since interior points of $\gamma^k \Delta$ come from interior points of Δ (before reducing modulo 2π, γ^k is an isomorphism), we just have to look at the points on the boundary. The proof will be completed once we prove that any point on $\partial\Delta$ comes from some point in $\Delta \cap \partial\Delta$ (unique by what we have already proved) that is if $\varphi_j = 2\pi l_j/l_{j-1}$ for $j \in J = \{j_1, \ldots, j_s\}$, $0 \le \varphi_j < 2\pi l_j/l_{j-1}$ for $j \notin J$, then $\varphi = \gamma^k \tilde{\varphi}$ with $\tilde{\varphi}_j = 0$ for j_1 and $0 \le \tilde{\varphi}_j < 2\pi l_j/l_{j-1}$ for the others. Of course for $j \in J$, we may assume that $l_j < l_{j-1}$.

For $2 \leq s \leq m$, $\gamma^k \varphi$ with k a multiple $\tilde{k}$ of p/l_{s-1} is such that $m_j \tilde{k} p/p l_{s-1} = (m_j/l_{s-1})\tilde{k}$, i.e., this is an integer for $j \leq s-1$ and these components are kept fixed, while for $j = s$, m_s and l_{s-1} have l_s as l.c.d. Then if $l_{s-1} = \alpha l_s, m_s = \beta l_s$, α and β are coprime and $e^{i 2\pi \tilde{k}\beta/\alpha}$ has α points in its orbit. But $k = \tilde{k}p/l_{s-1} < p/l_m$ so that $\tilde{k} < (l_{s-1}/l_s)(l_s/l_m)$. Thus there are l_s/l_m points $(k's)$ such that $\varphi_s = 2\pi l_s/l_{s-1} = 2\pi/\alpha$ is transformed into 0. Notice also that if φ_s is arbitrary in $\mathbb{R}$ then there is a $\tilde{k}$ (unique if $\tilde{k} \leq \alpha$) such that φ_s is sent into the interval $0 \leq \varphi_s < 2\pi l_s/l_{s-1}$.

Thus apply such a γ^k for j_1, keeping φ_j fixed $[mod.2\pi]$ for $j < j_1$ and sending $2\pi l_j/l_{j-1}$ into 0. Of course the other phases for $j > j_1$ will be moved. For $j_1 + 1$ apply again such a γ^k, keeping φ_j fixed for $j \leq j_1$ and sending φ_{j_1+1} in the interval $(0, 2\pi l_{j_1+1}/l_{j_1})$ and so on up to φ_m.

$$\text{Q.E.D.}$$

Remark 2.1. Note that if $\gamma^k \varphi$ fixes $\varphi_1, \ldots, \varphi_{s-1}$, then k is a multiple of p/l_{s-1} and then one gets a cyclic group of order l_{s-1}/l_m. Replacing k by $\tilde{k}p/l_{s-1}$ and p by l_{s-1} in the Lemma, one gets that the l_{s-1}/m_0 images of $\Delta_s = \{(0,\ldots,0,\varphi_s,\ldots,\varphi_m), 0 \leq \varphi_j < 2\pi l_j/l_{j-1}, j = s, \ldots, m\}$ with $\tilde{k} = 0, \ldots, l_{s-1}/m_0 - 1$, cover properly the set $\varphi_j = 0, j = 1, \ldots, s-1, \varphi_l \in [0, 2\pi[, l = s, \ldots, m$. In particular, we have that the face of $\bar{\Delta}$ with $\varphi_m = 2\pi l_m/l_{m-1}$ is sent directly (i.e., with φ_j fixed $j < m$) onto the base $\varphi_m = 0$. Thus the vertex $(0, \ldots, 0, 2\pi l_m/l_{m-1})$ is an image of $(0, \ldots, 0)$ and then, by the uniqueness, we can conclude that on the edge joining these two vertices there are no other images of $(0, 0, \ldots, 0)$.

Let S^1 act on $V = \mathbb{R} \times \mathbb{R}^M$ and $W = \mathbb{R} \times \mathbb{R}^N$ by $e^{im\varphi}z$ and $e^{in\varphi}z$, respectively as described in Chapter one. As usual we shall order the isotropy subgroups of S^1 in a decreasing order:

$$S^1, \mathbb{Z}_p, \mathbb{Z}_q, \ldots, \{e\}, \quad q < p$$

(cfr.[**t.D**, page 203]).

Let $H = \mathbb{Z}_{m_0}$ be an isotropy subgroup and let V^H be its fixed point subspace (i.e., $H = \{0, 1, \ldots, m_0 - 1\}$ and the action is given by $e^{i\varphi}$, $\varphi = 2k\pi/m_0$ with $k = 0, \ldots, m_0 - 1$). Then a point $(t, x_0, z_1, \ldots, z_m)$ belongs to V^H if and only if $z_j = 0$ or m_j is a multiple of

m_0. All components z_j with m_j multiple of m_0 appear in V^H and, since H is an isotropy subgroup, m_0 is the largest common divisor of those $m'_j s$.

Let B be the cylinder,

$$B = \{(t, x_0, z) \in \mathbb{R}^{M+1} : 0 \le t \le 1, |x_0| \le R, \|z\| \le R\}$$

where $\|.\|$ denotes the maximum norm, and let S be the boundary of B. Moreover, we set $B^H = B \cap V^H$ and $S^H = S \cap V^H$ for any isotropy subgroup H.

If $f : V \to W$ is a S^1-equivariant map, then we denote by f^H its restriction to V^H for any isotropy subgroup H. Note that $f^H : V^H \to W^H$ since $f(e^{i\varphi}(t, x_0, z)) = e^{i\varphi} f(t, x_0, z) = f(t, x_0, z)$ for $\varphi \in H$, and that f^H is S^1-equivariant. Thus $f = (f_0, F_0, \ldots, F_j, \ldots)$ with $F_j|_{V^H} = 0$ for those $j's$ such that n_j is not a multiple of m_0.

Let $H < K$ be isotropy subgroups with $H = \mathbb{Z}_{m_0}$. Then either $K = S^1$ or $K = \mathbb{Z}_{qm_0}$, with $q > 1$. Moreover, if K has the fixed point subspace of largest dimension among all isotropy subgroups such that $H < K$, then a point $(t, x_0, z) \in V^H \setminus V^K$ $(V^K \subset V^H)$ is such that z has a non zero component z_p with action $e^{ip\varphi}$ where p/m_0 and q are coprime. In fact, if p/m_0 and q have a common factor α then the subspace generated by V^K and z_p is contained in $V^{\tilde{K}}$ with $\tilde{K} = \mathbb{Z}_{\alpha m_0}$. Thus, $V^K \subsetneq V^{\tilde{K}} \subsetneq V^H$ contradicting the maximality of the dimension of V^K.

2.2. The Extension Theorem.

We are now in the position to state one of the main tools in the computation of the S^1-degree.

Theorem 2.1. *If $dim_{\mathbb{R}} V^H \le dim_{\mathbb{R}} W^H$, then any S^1-equivariant map* $f : \partial B \to W \setminus \{0\}$ *which has S^1-extensions $f^K : B^K \to W^K \setminus \{0\}$ for all $K's$ with $H < K$, agreeing on intersections, has a S^1-extension $\tilde{f} : B^H \to W^H \setminus \{0\}$. In particular, the map $\tilde{f}$ extends each f^K, too.*

Remark 2.2. At first note that S^1 acts on $V^H \setminus \cup_{H<K} V^K$ with just one isotropy orbit type. Thus one may extend, by the Tietze-Gleason Lemma, the maps f^K to neighborhoods B_ϵ^K of B^K in V^H with a non zero equivariant map $\hat{f}$. On the complement of $\cup B_\epsilon^K$ the

group S^1/H acts freely (cfr.[**B**, page 90]). Moreover, from [**B**, Corollary 5.12, page 89], we know that S^1-equivariant maps from $B^H \setminus \cup_{H<K} B_\epsilon^K$ into $W^H \setminus \{0\}$ are in one to one correspondence with S^1/H-equivariant maps between the same spaces. Finally, from [**t.D**, Proposition 8.3.1, page 211] (cfr. also [**B**, Theorem 5.13, page 90 and Theorem 9.3, page 105]) the dimension of $(B^H \setminus \cup_{H<K} B_\epsilon^K)/S^1$ in strictly less then the dimension of S^H, thus one may extend the map $\hat{f}$ to a S^1-equivariant map from B^H into $W^H \setminus \{0\}$.

However this result depends on obstruction theory and one of the goals of this work is to circumvent this theory as much as possible. Also we will need a direct proof of it in order to compute the *extension degree* which we will define below. What we shall do below is to give an explicit equivariant cell decomposition for the above sets.

Proof of the Extension Theorem. The proof will be by induction on $dim_{\mathbb{R}} V^H$, keeping fixed k, l, W^H.

Let $dim_{\mathbb{R}} V^H = (k+1) + 2$ (the lowest admissible case). If $H < K$, then the subgroup K is S^1. This case was treated in [**I.M.V**, Theorem 4.4 (i)], and, keeping the same notation, it reads as follows:

$$dim_{\mathbb{R}} V^H = (k+1) + 2, m = 1, 2n = dim_{\mathbb{R}} W^H - l - 1$$

and the condition $dim_{\mathbb{R}} V^H \leq dim_{\mathbb{R}} W^H$ implies

$$k + 1 + 2m \leq l + 1 + 2n \quad \text{that} \quad \text{is} \quad m < h_c.$$

By Theorem 1.1 (i), we have the assertion.

The next case is $dim_{\mathbb{R}} V^H = (k+1) + 4$, $m = 2$. If the only K with $H < K$ is $K = S^1$ then again Theorem 1.1 applies. Thus one has to consider K with $K = \mathbb{Z}_{q \, m_0}$ and $dim_{\mathbb{R}} V^K = (k+1) + 2$.

Writing the elements of V^H as (t, x_0, z_1, z_2), then the elements of V^K will be of the form $(t, x_0, 0, z_2)$ with $m_1 = q_0 \, m_0$, $m_2 = q \, m_0$, $q > q_0$, q and q_0 coprime. (For notational purposes we omit the other variables z_j with m_j not a multiple of m_0).

a) If $q_0 = 1$, consider, on the boundary of the ball

$$\{0 \le t \le 1, |x_0| \le R, |z_2| \le R, \quad z_1 \in \mathbb{R}, 0 \le z_1 \le R\},$$

the map $\bar{f}$ defined as

$$\bar{f} = \begin{cases} f^K & \text{for } z_1 = 0 \\ f & \text{for } z_1 > 0. \end{cases}$$

Then $\bar{f}$ defines an element in $\Pi_{k+3}(S^{l+2n})$. But, from the dimension assumption, we have that $\Pi_{k+3}(S^{l+2n}) = 0$ so that the map $\bar{f}$ has a non zero continuous extension $\tilde{f}$ to the ball.

Now, for $z_1 = |z_1| e^{i\psi}$ define

$$f^H(t, x_0, z_1, z_2) = e^{i\psi/m_0} \tilde{f}(t, x_0, |z_1|, e^{-iq\psi} z_2) =$$
$$= (\tilde{f}_0, \tilde{F}_0, \ldots, e^{i\psi n_j/m_0} \tilde{F}_j, \ldots)(t, x_0, |z_1|, e^{-iq\psi} z_2) \ .$$

Recall that if n_j is not a multiple of m_0, then the component $\tilde{F}_j$ is not in W^H and thus it is 0. Furthermore, observe that there is no indeterminacy in the phase: ψ is defined modulo 2π.

If $z_1 = 0$, then, by the equivariance of f^K, we have

$$e^{i\psi/m_0} \tilde{f}(t, x_0, 0, e^{-iq\psi} z_2) = e^{i\psi/m_0} f^K(t, x_0, 0, e^{im_2(-\psi/m_0)} z_2)$$
$$= e^{i\psi/m_0} f^K(t, x_0, e^{i(-\psi/m_0)} z)$$
$$= f^K(t, x_0, z),$$

i.e., f^H is well-defined. (One can avoid this last step by using Gleason's Lemma in order to define f^K for $|z_1| \le \epsilon$.)

Moreover, the map f^H is S^1-equivariant since, setting $z_1 = e^{i\psi} |z_1|$, we have

$$f^H(t, x_0, e^{i\varphi} z) = f^H(t, x_0, e^{im_1\varphi} z_1, e^{im_2\varphi} z_2),$$
$$= e^{i(\psi + m_1\varphi)/m_0} \tilde{f}(t, x_0, |z_1|, e^{-iq\psi - iqm_1\varphi} e^{im_2\varphi} z_2)$$
$$= e^{i\varphi} e^{i\psi/m_0} \tilde{f}(t, x_0, |z_1|, e^{-iq\psi} z_2)$$
$$= e^{i\varphi} f^H(t, x_0, z_1, z_2).$$

This proves the result for this case.

b) If $q_0 > 1$, since q and q_0 are coprime, one has also a S^1-extension $f^{\tilde{K}}$ to $B^{\tilde{K}}$, where $\tilde{K} = \mathbb{Z}_{q_0 m_0}$ and $V^{\tilde{K}} = \{(t, x_0, z_1, 0)\}$. By Gleason's Lemma, we may assume that f^K is S^1-equivariant and non zero for $|z_1| \leq \epsilon$ and, as well, $f^{\tilde{K}}$ for $|z_2| \leq \epsilon$ (agreeing on the intersection). We need a nonvanishing S^1-extension to the set

$$\{t \in [0,1], \|x_0\| \leq R, \epsilon \leq |z_1| \leq R, \epsilon \leq |z_2| \leq R\}.$$

Consider the set

$$D = \{t \in [0,1], \|x_0\| \leq R, z_1, z_2 \in \mathbb{R}, \epsilon \leq z_1 \leq R, \epsilon \leq z_2 \leq R\},$$

a ball of dimension $k+3$. On ∂D the map defined by f, f^K for $z_1 = \epsilon$ and $f^{\tilde{K}}$ for $z_2 = \epsilon$ is non zero and defines an element of $\Pi_{k+2}(S^{l+2n}) = 0$. Thus one has an extension $\tilde{f}$ to the full ball D.

Consider now the ball of dimension $k+4$

$$D' = \{t \in [0,1], \|x_0\| \leq R, z_1 \in \mathbb{R}, \epsilon \leq z_1 \leq R,$$
$$z_2 = |z_2|e^{i\varphi_2}, \epsilon \leq |z_2| \leq R, 0 \leq \varphi_2 \leq 2\pi m_0/m_1 = 2\pi/q_0\}.$$
$$\Delta = \{\varphi_2, 0 \leq \varphi_2 = 2\pi/q_0\}.$$

Moreover, let k_0 be the unique integer, $1 \leq k_0 < q_0$, such that

$$e^{2\pi i m_2 k_0/m_1} e^{-2\pi i/q_0} = 1$$

(see the Fundamental Cell Lemma). Define a map

$$\tilde{f}_{k_0}(t, x_0, |z_1|, |z_2|e^{2\pi i/q_0}) := e^{2\pi i k_0/m_1} \tilde{f}(t, x_0, |z_1|, |z_2|) =$$
$$= (\tilde{f}_0, \tilde{F}_0, \ldots, e^{2\pi i k_0 n_j/m_1} \tilde{F}_j, \ldots)(t, x_0, |z_1|, |z_2|) \ .$$

Since the maps f, f^K, $f^{\tilde{K}}$ have the property that

$$f(t, x_0, e^{2\pi i m_1 k_0/m_1}|z_1|, e^{2\pi i m_2 k_0/m_1}|z_2|) =$$
$$= e^{2\pi i k_0/m_1} f(t, x_0, |z_1|, |z_2|),$$

the map $\tilde{f}_{k_0}$ connects continuously with the functions f, f^K, $f^{\tilde{K}}$ and $\tilde{f}$. Note that the map f has no other symmetries keeping fixed the phase of z_1.

Now, consider on $\partial D'$ the continuous map given by f, f^K, $f^{\tilde{K}}$, $\tilde{f}$ for $\varphi_2 = 0$ and $\tilde{f}_{k_0}$ for $\varphi_2 = 2\pi/q_0$. We have then a non zero continuous function which defines an element of $\Pi_{k+3}(S^{l+2n}) = 0$, hence it has an extension, that we still denote by $\tilde{f}$, to the $(k+4)$-dimensional ball D'.

We shall define non zero extensions to the cells $\gamma^k \Delta, k = 1, \ldots, q_0 - 1$, by defining

$$\tilde{f}_k(t, x_0, |z_1|, \gamma^k z_2) := e^{2\pi i k/m_1} \tilde{f}(t, x_0, |z_1|, z_2) \text{ for } z_2/|z_2| \text{ in } \Delta.$$

Since $f, f^K, f^{\tilde{K}}, \tilde{f}_{k_0}$ satisfy the symmetry

$$f(t, x_0, e^{2\pi i m_1 k/m_1}|z_1|, e^{2\pi i m_2 k/m_1} z_2) = e^{2\pi i k/m_1} f(t, x_0, |z_1|, z_2)$$

we shall get a continuous non zero map, called also $\tilde{f}$, on the set

$$\{t \in [0,1], \|x_0\| \leq R, \epsilon \leq z_1 \leq R, \epsilon \leq |z_2| \leq R\}.$$

For $z_1 = e^{i\psi}|z_1|$ define

$$f^H(t, x_0, z_1, z_2) := e^{i\psi/m_1} \tilde{f}(t, x_0, |z_1|, e^{-i m_2 \psi/m_1} z_2).$$

If ψ is replaced by $\psi + 2k\pi$, then by the construction of $\tilde{f}$ we have that

$$e^{i(\psi + 2\pi k)/m_1} \tilde{f}(t, x_0, |z_1|, e^{-i m_2 \psi/m_1} e^{-2\pi i k m_2/m_1} z_2) =$$
$$= e^{i\psi/m_1} \tilde{f}(t, x_0, |z_1|, e^{-i m_2 \psi/m_1} z_2).$$

Thus the map f^H is well defined. Furthermore,

$$f^H(t, x_0, e^{i m_1 \varphi} z_1, e^{i m_2 \varphi} z_2) = e^{i(\psi/m_1 + \varphi)} \tilde{f}(t, x_0, |z_1|, e^{-i(m_2\psi/m_1 + m_2\varphi)} e^{i m_2 \varphi} z_2)$$
$$= e^{i\varphi} f^H(t, x_0, z_1, z_2).$$

Thus, one has the result for $m = 2$. The induction step is then as follows.

Assume that the result holds if

$$dim_{\mathbb{R}} V^H = (k+1) + 2m - 2 \leq (l+1) + 2n - 2.$$

Then we shall prove it if $dim_{\mathbb{R}} V^H = (k+1) + 2m$.

Denote by z_p the component of $V^H \backslash V^K$ with corresponding action $e^{ip\varphi}$ with lowest p (K isotropy subgroup with $H < K$, V^K of maximal dimension). If $K = S^1$, then Theorem 1.1 applies. Let $\hat{z} = (z_2, \ldots, z_m)$ be the other components of V^H. Let $q\, m_0$ be the largest common divisor of $m_2, \ldots, m_m$ (q may be 1). Then p/m_0 and q are coprime.

a) If $p = m_0$ then, by the inductive hypothesis on $V^H \cap \{z_p = 0\}$, one has an equivariant extension of the maps defined on $S^H \cap \{z_p = 0\} \cap \bigcup_{H < K} B^K$ to the set $B^H \cap \{z_p = 0\}$.

Consider, on the boundary of the ball,

$$\{t \in [0,1], \|x_0\| \leq R, |\hat{z}| \leq R, 0 \leq z_p \leq R\}$$

the map defined as f if $z_p > 0$ and, for $z_p = 0$, as the equivariant extension given by the inductive hypothesis. This map defines an element of $\Pi_{k+2m-1}(S^{l+2n}) = 0$. Hence it has a continuous extension to the full ball.

For $z_p = e^{i\psi}|z_p|$, define

$$f^H(t, x_0, z_p, \hat{z}) := e^{i\psi/m_0} \hat{f}(t, x_0, |z_p|, e^{-i\psi/m_0}\hat{z})$$

where $e^{-i\psi/m_0}\hat{z} := (e^{-im_2\psi/m_0}z_2, \ldots, e^{-im_m\psi/m_0}z_m)$. As before, $\tilde{F}_j \equiv 0$ for n_j not a multiple of m_0 and the $m_j's$ are all multiples of m_0, thus there is no indeterminacy in the choice (mod. 2π) of ψ. If $z_p = 0$,

$$\hat{f}(t, x_0, 0, e^{-i\psi/m_0}\hat{z}) = e^{-i\psi/m_0}\hat{f}(t, x_0, 0, \hat{z})$$

by the equivariance of the map obtained by the inductive hypothesis, thus f^H is well defined. Furthermore,

$$f^H(t, x_0, e^{i\varphi}z) = e^{i(\psi+p\varphi)/m_0}\hat{f}(t, x_0, |z_p|, e^{-i(\psi+p\varphi)/m_0}e^{i\varphi}\hat{z}) =$$
$$= e^{i\varphi}f^H(t, x_0, z).$$

b) If $p > m_0$, then by the inductive hypothesis there is a S^1-extension to $B^H \cap \{z_p = 0\}$. Before applying the Fundamental Cell Lemma, order the vector $(z_2, \ldots, z_m) = (\tilde{z}, \hat{z})$ by setting in $\hat{z}$ all the components z_j such that $l_j = l_{j-1}$. One may reduce the size of $\tilde{z}$ by denoting as z_2 the component with the smallest l.c.d. l_2 with p. Put all z's with m_j a multiple of l_2 in $\hat{z}$. If $l_2 > m_0$, then there are $z's$ left out (if not the l.c.d. of $p, m_2, \ldots, m_m$ would be at least l_2). Among these $z's$ denote by z_3 the component z such that the l.c.d. l_3 of (p, m_2, m_3) is smallest. Put in $\hat{z}$ all $z's$ with m_j a multiple of l_3. And so on.

Let $(z_2, \ldots, z_r) = \tilde{z}$. By the inductive hypothesis there is a S^1-extension to

$$\bigcup_{j=p,2,\ldots,r} B^H \cap \{z_j = 0\}$$

and by Gleason's Lemma to an ϵ-neighborhood of this set. We need to extend the map equivariantly to the set $S =: \{t \in [0,1], \|x_0\| \leq R, \epsilon \leq |z_p| \leq R, \epsilon \leq |z_j| \leq R, j = 2, \ldots, r; |z_j| \leq R, j > r\}$.

On this set the action of S^1 has just one isotropy type $\mathbb{Z}_{m_0}$. Consider the ball (subset of the previous set) $C := \{t \in [0,1], \|x_0\| \leq R, |\hat{z}| \leq R, \epsilon \leq z_p \leq R, \epsilon \leq |z_j| \leq R, \varphi = (\varphi_2, \ldots, \varphi_r), 0 \leq \varphi_j < 2\pi l_j/l_{j-1}, i.e., \varphi \in \Delta\}$. Clearly, C has dimension $k+1+2m-1$.

From the Fundamental Cell Lemma, $\{\gamma^k C\}$, $k = 0, \ldots, p/m_0 - 1$, cover properly $S \cap \{z_p > 0\}$, when $\{\gamma^k\}, k = 0, \ldots, p-1$, is the isotropy subgroup $\mathbb{Z}_p$ of the component z_p. The equivariant extension on S will be constructed from a $\mathbb{Z}_p$-equivariant extension on $S \cap \{z_p > 0\}$, coming from a continuous extension on C. However this last extension comes from a map defined on ∂C which has to be compatible with the action of $\mathbb{Z}_p$ on the walls of Δ (as seen in the Fundamental Cell Lemma).

The extension on ∂C has to be done step by step, in such a way that it preserves symmetries and is compatible with the previous extensions. This process will done in the following way.

If $\varphi_j = 0$, $j = 2, \ldots, r$, one obtains a $k + 1 + 2m - r$ ball, with a non zero map defined on its boundary. Since $2 \leq r$, $\Pi_{k+2m-r}(S^{l+2n}) = 0$ and one gets a continuous non zero extension on the vertex of Δ, $\varphi = 0$. Extend this map to the p/m_0 images of the vertex under the action of $\mathbb{Z}_p$:

$$\tilde{f}(t, x_0, \hat{z}, |z_p|, |z_2|, \ldots, |z_r|, \gamma^k(1, \ldots, 1)) := e^{i2\pi k/p} \tilde{f}(t, x_0, \hat{z}, |z_p|, |z_2|, \ldots, |z_r|, (1, \ldots, 1)),$$

where we are representing $\tilde{z}$ in polar coordinates and the vector $(1,\ldots,1)$ corresponds to $\varphi = 0$. Since the argument of z_p is fixed under $\mathbb{Z}_p$ (and only under this group or subgroups) this extension is coherent with the equivariance of f.

Let $\Delta_r = (0,\ldots,0,\varphi_r)$, $\Delta_{r-1} = (0,\ldots,0,\varphi_{r-1},\varphi_r),\ldots,\Delta_2 = \Delta$.

Consider now the ball

$$C_r = \{t \in [0,1], \|x_0\| \leq R, \|\hat{z}\| \leq R, \epsilon \leq z_p \leq R, \epsilon \leq |z_j| \leq R, \varphi \in \Delta_r\}$$

and on ∂C_r, the map defined by the previous extensions (for $\varphi_r = 2\pi l_r/l_{r-1}$ this extension is given in terms of the vertex of Δ). Since $dim\ \partial C_r = k+1+2m-r < l+2n$, one obtains a continuous extension to C_r, which will be reproduced by γ^k to the other one-dimensional images of Δ_r. From Remark 2.1, C_r doesn't contain any other image of $(0,\ldots,0)$.

For $\varphi \in \Delta_{r-1}$, we have already an extension for $\varphi \in \bar{\Delta}_r$. For $\varphi_0 = (0,\ldots,$ $2\pi l_{r-1}/l_{r-2},0)$, we define the extension from γ^k with $\gamma^k \varphi_0 \in \bar{\Delta}_r$. For $\varphi_r = 0$, $\varphi_{r-1} \in [0, 2\pi l_{r-1}/l_{r-2}]$ one has a problem as above, hence one gets a continuous extension (there is no problem of compatibility from the uniqueness in the Fundamental Cell Lemma: no points in this set (except $\varphi_{r-1} = 2\pi l_{r-1}/l_{r-2}$) come from $\bar{\Delta}_r$). For $\varphi_r \in [0, 2\pi l_r/l_{r-1}]$, $\varphi_{r-1} = 2\pi l_{r-1}/l_{r-2}$, extend from γ^k with $\gamma^k \varphi \in \bar{\Delta}_r$ (one may have several k's for this line, but again one has continuity and compatibility at each point. Take the example : $p = 12$, $m_1 = 10$, $m_2 = 15$). The last part of the boundary of $\bar{\Delta}_{r-1}$, $\varphi_r = 2\pi l_r/l_{r-1}$, $0 \leq \varphi_{r-1} \leq 2\pi l_{r-1}/l_{r-2}$ is given from the extension for $\varphi_r = 0$ with γ^k which leaves φ_{r-1} unchanged. One obtains then a continuous extension to Δ_{r-1}, since $k+1+2m-r+1 < l+2n$. Note that if γ^k fixes $\varphi_1,\ldots,\varphi_{r-2}$ then one doesn't obtain any other symmetries than the ones used in the above extension.

Assume now that we have an extension on $\bar{\Delta}_{s+1}$ with the symmetries given by γ^k on $\bar{\Delta}_{s+1} \setminus \Delta_{s+1}$, that is, k corresponds to the unique subgroup which fixes $\varphi_1,\ldots,\varphi_s$. The images of Δ_{s+1} under this subgroup cover properly all the possible phases $\varphi_{s+1},\ldots,\varphi_r$ with $(\varphi_1,\ldots,\varphi_s) = 0$ (see Remark 2.1).

Consider $\Delta_s = (0,\ldots,0,\varphi_s,\ldots,\varphi_r)$. On Δ_s one has to consider only the symmetries given by γ^k with k a multiple of p/l_{s-1}, which leave unchanged $\varphi_1,\ldots,\varphi_{s-1}$. The Fundamental Cell Lemma applies with p replaced by l_{s-1} and $j \geq s$ (l_j is the same).

The face of Δ_s with $\varphi_s = 0$ is $\bar{\Delta}_{s+1}$. For the face of $\bar{\Delta}_s$ with $\varphi_s = 2\pi l_s/l_{s-1}$ apply γ^k with $\gamma^k(2\pi l_s/l_{s-1}) = 0$ and define the extension from the images of that face on $\bar{\Delta}_{s+1}$. For $0 < \varphi_s < 2\pi l_s/l_{s-1}$, $\gamma^k\varphi$ will be out of Δ_s unless, as explained in the proof of the Fundamental Cell Lemma, k is a multiple of p/l_s, that is for the subgroup acting on Δ_{s+1} and fixing $\varphi_1, \ldots, \varphi_s$. One constructs first the extension for $0 \le \varphi_s \le 2\pi l_s/l_{s-1}$, $\varphi_j = 0$ $j \ne s$, by extending the values on the boundary to the interval (the interior of the interval is in Δ and no point on it comes from Δ_{s+1}). On the boundary of

$$\{0 \le \varphi_r \le 2\pi l_r/l_{r-1}, 0 \le \varphi_s \le 2\pi l_s/l_{s-1}, \varphi_j = 0, j \ne r, s\}$$

one has the previous extensions for $\varphi_r = 0$, $\varphi_s = 0$, $\varphi_s = 2\pi l_s/l_{s-1}$ (given in terms of $\bar{\Delta}_{s+1}$) and for $\varphi_r = 2\pi l_r/l_{r-1}$ in terms of the values for $\varphi_r = 0$ (this γ^k leaves unchanged all the other coordinates by the Fundamental Cell Lemma with p replaced by l_s, the interior of this square is not affected by the action of the group).

If one has extended the map to

$$\{0 \le \varphi_s \le 2\pi l_s/l_{s-1}, \varphi_j = 0, s \ne j \le s+t, 0 \le \varphi_j \le 2\pi l_j/l_{j-1}, s+t+1 \le j\}$$

adding the component $\varphi_{s+t} \in [0, 2\pi l_{s+t}/l_{s+t-1}]$, one gets the value of the map for $\varphi_{s+t} = 2\pi l_{s+t}/l_{s+t-1}$ by acting on φ_j with $j \ge s+t$, leaving fixed φ_j, $j < s+t$.

For $\varphi_j = 0$, $j \ne s, s+t$, $0 \le \varphi_s \le 2\pi l_s/l_{s-1}$, $0 \le \varphi_{s+t} \le 2\pi l_{s+t}/l_{s+t-1}$, one knows the function on its boundary and one may extend it to the ball by the dimension argument. Taking the same set crossed with $0 \le \varphi_r \le 2\pi l_r/l_{r-1}$, one has again a known value on its boundary (using γ^k) and, by dimension, on the set itself.

A new induction argument will reduce the problem to the extension on the set

$$\{0 \le \varphi_s \le 2\pi l_s/l_{s-1}, 0 \le \varphi_{s+t} \le 2\pi l_{s+t}/l_{s+t-1}, 0 \le \varphi_{s+t_1} \le 2\pi l_{s+t_1}/l_{s+t_1-1}, t_1 > t,$$
$$\varphi_j \in [0, 2\pi l_j/l_{j-1}), j > t\}.$$

One starts with the boundary of the set $\{\varphi_s, \varphi_{s+t}, \varphi_{s+t_1}, \varphi_j = 0$,   for $j \ne s, s+t, s+t_1\}$ then with the ball crossed with the interval for φ_r and so on. This chain of induction arguments will end (since $t_1 < t_2 < \cdots \le r$), proving all inductions. That is once one knows how to extend on the boundary by a map which is consistent with the action of γ^k, (recall that each point on the boundary of $\bar{\Delta}_s \backslash \Delta_s$ comes from a unique point in Δ_s, with the action of γ^k (k unique modulo l_s/m_0) and that this γ^k sends the face with lowest index not in Δ_s and moves only coordinates with higher indices. Thus one gets the equivariant map on the boundaries of the different cubes by a descending order of indices and inside these cubes by a dimension argument), one extends inside the ball by the dimension argument

Indeed, if $C_s = C \cap \{\varphi_j = 0; j = 2, \ldots, s - 1\}$, then

$$dim C_s = (k + 1) + 2m - s + 1 < (k + 1) + 2m \leq (l + 1) + 2n,$$

since $2 \leq s$.

Once the extension to the ball C is performed, one extends for $\varphi \in [0, 2\pi]^{r-1}$ by using the action of the group γ^k, namely:

$$\tilde{f}(t, x_0, |z_p|, \hat{z}, \gamma^k \tilde{z}) = e^{2\pi i k/p} \tilde{f}(t, x_0, |z_p|, e^{-2\pi i k/p} \hat{z}, \tilde{z}) \quad \text{for } \tilde{z}/|\tilde{z}| \in \Delta.$$

One extends finally $\tilde{f}$ to $z_p = |z_p| e^{i\psi}$ by defining

$$f^H(t, x_0, z_p, \hat{z}, \tilde{z}) = e^{i\psi/p} \tilde{f}(t, x_0, |z_p|, e^{-i\psi/p} \hat{z}, e^{-i\psi/p} \tilde{z}).$$

If ψ is replaced by $\psi + 2\pi k$, then

$$e^{i\psi/p} e^{2\pi i k/p} \tilde{f}(t, x_0, |z_p|, e^{-i\frac{\psi}{p} - i\frac{2k\pi}{p}} \hat{z}, e^{-i\frac{\psi}{p} - i\frac{2k\pi}{p}} \tilde{z}) = e^{i\psi/p} \tilde{f}(t, x_0, |z_p|, e^{-i\psi/p} \hat{z}, e^{-i\psi/p} \tilde{z})$$

by the construction of $\tilde{f}$, thus f^H is well defined. Furthermore

$$f^H(t, x_0, e^{i\varphi} z) = e^{i(\varphi + \psi/p)} \tilde{f}(t, x_0, |z_p|, e^{-i\psi/p} e^{i\varphi} e^{-i\varphi} (\hat{z}, \tilde{z}))$$
$$= e^{i\varphi} \tilde{f}(t, x_0, z).$$

Q.E.D.

Note that since $\mathbb{Z}_p$ is the maximal group which leaves z_p real and positive then this construction is compatible with the previous S^1-maps.

Corollary 2.1. *If for all isotropy subgroups H of S^1 one has that $dim V^H \leq dim W^H$, then a S^1-map $f : S \longrightarrow W \setminus \{0\}$ has a non zero S^1-extension if and only if $f : S^k \longrightarrow \mathbb{R}^{l+1} \setminus \{0\}$ is trivial. Thus, if $k < l$, $\Pi^{S^1}_{k+2m}(S^{l+2n}) = 0$.*

Proof. The dimension hypothesis implies that $k \le l$, $k + 2m \le l + 2n$. Thus the result is a direct consequence of the extension theorem.

Q.E.D.

Corollary 2.2 *Under the same assumptions of Corollary 2.1, if $k = l$ and if there exists an equivariant map $F : \mathbb{C}^m \setminus \{0\} \longrightarrow \mathbb{C}^n \setminus \{0\}$, then $\Pi_{k+2m}^{S^1}(S^{l+2n}) = \mathbb{Z} = \Pi_k(S^l)$.*

Furthermore, if $m = n$, then a S^1-map $(f_0(t, x_0, z), h(t, x_0, z)) : S \to W \setminus \{0\}$ has an usual Brouwer degree equal to the product of the index of the map F with the degree of $f_0(t, x_0, 0)$ with respect to $I \times \{x_0 : \|x_0\| \le R_0\}$. The index of F is independent of the choice of F, that is any other choice of F will give the same index.

Proof. In fact, if $f(t, x_0)$ is an element of $\Pi_k(S^l)$ then one may extend this map to $I \times B_0$ radially and in such a way that the only zero is at $t = 1/2$, $x_0 = 0$. Then the map defined as $f(t, x_0, z) = (f_0(t, x_0), F(z))$ will give the desired element in $\Pi_{k+2m}^{S^1}(S^{l+2n})$. Thus the morphism $P_* : \Pi_{k+2m}^{S^1}(S^{l+2n}) \longrightarrow \Pi_k(S^l)$ is onto. From Corollary 2.1, P_* is one to one.

The last point comes from the fact that (f_0, h) and $(f_0(t, x_0, 0), F(z))$ have the same image in $\Pi_k(S^l)$ under the morphism P_*. These two are thus S^1-homotopic and, a fortiori, are plainly homotopic. The independence from F is then clear.

Q.E.D.

Remark 2.3. Under the above dimensional conditions, the existence of such a map F seems to be always true, however to prove it would take us away from our main purpose. It is enough to remark that if $n_j = k_j m_j$ for $j = 1, \cdots, m$, then the dimension conditions are met and one can take as F the map defined by

$$F(z_1, \cdots, z_m) = (z_1^{k_1}, \cdots, z_m^{k_m}, 0, \cdots, 0),$$

where k_j are non-zero integers and z^k, for negative k, is meant to be $\bar{z}^{|k|}$. If $m = n$, this map F has index Πk_j. This example covers the case of the same action, i.e., $k_j = 1$ and the case studied in [**I.M.V**].

For example, if $m = n = 2$, $m_1 \le m_2$, denoting by m_0 the l.c.d. of m_1, m_2, then the dimension conditions imply that n_1 and n_2 are multiples of m_0. Thus, one of them (at least) is a multiple of m_2 (say n_2) and also one of them is a multiple of m_1. Thus, besides the case $n_j = k_j m_j$, one could have that n_2 is a multiple of m_1 and m_2 and n_1 is not. In this case if p and q are the integers such that $p m_1 + q m_2 = m_0$ (given by Darboux ' theorem), then one could take

$$F(z_1, z_2) = ((z_1^p z_2^q)^{n_1/m_0}, z_1^{n_2/m_1} + z_2^{n_2/m_2}).$$

2.3. The Extension degree.

We shall see now what are the obstructions for extending a map to B^H, assuming it has been extended to B^K, $H < K$, when

$$dim_{\mathbb{R}} V^H = dim_{\mathbb{R}} W^H + \delta, \ \delta = 1 \text{ or } 2.$$

Note that, since $k + 2m = l + 2n + \delta$, then $\delta = 1$ if and only if $|k - l|$ is odd and $\delta = 2$ if and only if $|k - l|$ is even.

Let z_p be, as in the proof of the Extension Theorem, the last component which does not belong to V^K, K being such that V^K is maximal.

a) If $p = m_0$, consider the ball

$$D = \{0 \le t \le 1, |x_0| \le R, 0 \le z_p \le R, |\tilde{z}_p| \le R\}.$$

On ∂D, one has an extension for $z_p > 0$ (just by taking the original map f) and for $z_p = 0$ by the dimension argument. Then this map defines an element of

$$\Pi_{k+2m-1}(S^{l+2n}) = \Pi_{l+2n+\delta-1}(S^{l+2n}) = \mathbb{Z} \text{ if } \delta = 1$$
$$= \mathbb{Z}_2 \text{ if } \delta = 2 \text{ and } 3 \le l + 2n$$

(the case $l + 2n < 3$ could be treated as in [**I.M.V.**]). This element will be called the *Extension degree of* f to V^H, and it will be denoted by $deg_E(f)$.

If $deg_E(f) = 0$, then one may extend the map to $z_p \in \mathbb{C}$ as before. We shall see that it does not depend on the previous extensions f^K.

b) If $p > m_0$, $\delta = 1$, consider the set

$$D = \{0 \le t \le 1, |x_0| \le R, \varepsilon \le z_p \le R, |\hat{z}| \le R, \varepsilon \le |z_j| \le R \, ,$$

$$j = 2, \cdots, r, \varphi = (\varphi_2, \cdots, \varphi_r) \in \bar{\Delta}\}.$$

On ∂D one has a continuous nonvanishing map, defined by the previous extensions and constructed as in the Extension Theorem: the walls of D have dimension $k + 2m - 1 = l + 2n + \delta - 1$, with boundary of dimension $k + 2m - 2 < l + 2n$, thus there is no problem for the extension on the walls of Δ. On the boundary of D, one has an element of $\Pi_{k+2m-1}(S^{l+2n}) = \mathbb{Z}$ which will be called $deg_E(f)$.

c) If $p > m_0$ and $\delta = 2$, then for a wall of Δ (for example $\varphi_s = 0$) one has a ball of dimension $k + 2m - 1 = l + 2n + 1$. Thus the dimension argument does not work for the extension to the boundary of D. One has to extend to the walls of Δ in the ordered way indicated in the extension theorem.

In principle there is a first obstruction for the extension to the walls of Δ. If this is zero then one will get a new obstruction in

$$\Pi_{k+2m-1}(S^{l+2n}) = \mathbb{Z}_2, \quad \text{if } 3 \le l + 2n.$$

We conjecture that there is always an extension to the walls of Δ and that the obstruction in $\mathbb{Z}_2$ is independent of the previous extension (the arguments used below for the case $\delta = 1$ do not work). Since the amount of work necessary to prove this result, for $\delta = 2$, is an order of magnitude higher than for the case $\delta = 1$, and since we do not have interesting applications at hand, we prefer to leave the case $\delta = 2$ as it stands and consider it in a future publication.

Lemma 2.2. ($\delta = 1$). *If $k \ne l - 1$, then the Extension degree is independent of the previous extensions. If $k = l - 1$, then the Extension degree is well defined modulo* $m_0(\Pi_1^{m-1}n_j)/(\Pi_1^m m_j)$.

Proof. We shall use the following detour in order to give a short proof of this result.

Let us consider the map $f(t, x_0, z_1, \cdots, z_m)$ on B^H, with its non-zero equivariant extension f^K to B^K for all $K > H$. Clearly, one will obtain through Θ^* non-zero extensions F^K for $F(t, X_0, Z_1, \cdots, Z_m) = \Theta^* f = f(t, X_0, Z_1^{\mu_1}, \cdots, Z_m^{\mu_m}), \mu_j = m_j/m_0$ (see Chapter 1). Let $\bar{F}$ be the map defined on the boundary of the ball

$\{0 \leq t \leq 1, |X_0| \leq R, 0 \leq Z_1 \leq R, |\bar{Z}_j| \leq R, j \geq 2\}$ as F for $Z_1 > 0$ and as $\Theta^* \tilde{f}$ for $Z_1 = 0$, where $\tilde{f}$ is the non-zero S^1-map obtained in the Extension Theorem and defining the extension degree. Recall that $deg_K(F)$ is the Brouwer degree of $\tilde{F}$ and is independent of the extension used for $Z_1 = 0$, if $k \neq l - 1$ and is well defined modulo $\Pi(n_j/m_0)$ if $k = l - 1$.

a) If $p = m_0$, i.e., $\mu_1 = 1$, then the map

$$(Z_2, \cdots, Z_m) \longmapsto (Z_2^{\mu_2}, \cdots, Z_m^{\mu_m})$$

has degree $\Pi_{i=2}^m \mu_i$ and, by the composition rule for the degree, one has

$$deg_K(\tilde{F}) = (\Pi_2^m \mu_i) deg_E(f) = \Pi_1^m (m_j/m_0) deg_E(f).$$

Thus, whenever $deg_K(\tilde{F})$ is independent of the previous extensions, so is the Extension degree.

b) If $p > m_0$, let D be the set

$$\{0 \leq t \leq 1, |x_0| \leq R, \varepsilon \leq z_p \leq R, |\hat{z}| \leq R, \varepsilon \leq |\tilde{z}_j| \leq R, \varphi \in \Delta\}$$

and $\gamma^k D$ be the image of this ball under the symmetry.

By the previous argument one has

$$deg_K(\tilde{F}) = \Pi_{i=2}^m \mu_i deg(f; \bigcup_{k=0}^{\mu_1-1} \gamma^k D \bigcup_j (B^H \bigcap \{|\tilde{z}_j| < \varepsilon\})).$$

Since the map f is non-zero on $B^H \bigcap \{|\tilde{z}_j| < \varepsilon\}$ by hypothesis, then by the excision property and by the additivity of the Brouwer degree one has that

$$deg_K(\tilde{F}) = (\Pi_2^m \mu_i) \sum_{k=0}^{\mu_1-1} deg(f; \gamma^k D).$$

However on $\gamma^k D$, one has that

$$f(t, x_0, |z_p|, \hat{z}, \gamma^k \tilde{z}) = e^{i2\pi k/p} f(t, x_0, |z_p|, e^{-2\pi ik/p} \hat{z}, \tilde{z})$$

and the maps $\hat{z} \longmapsto e^{-2\pi ik/p}\hat{z}$, $f \longmapsto e^{2\pi ik/p}f$ have degree 1, thus $deg(f; \gamma^k D) = deg(f; D)$. Hence

$$deg_K(\tilde{F}) = \Pi_1^m \mu_i deg(f; D) = \Pi_1^m (m_j/m_0) deg_E(f).$$

One has thus the same result and the extension degree is well defined.

If $k = l - 1$ and f_1, f_2 are two extensions, then, from Theorem 1.1 (iv), we have that $deg_K(\tilde{F}_1) - deg_K(\tilde{F}_2)$ is a multiple of $\Pi_1^{q_j=1}(n_j/m_0)$ where $2q+l+1 = dim_{\mathbb{R}} W^H$, $q = m-1$ in the above notation. Thus the difference of the two possible extension degrees would be a multiple of $\Pi_1^{m-1}(n_j/m_0)/\Pi_1^m(m_j/m_0) = m_0(\Pi_1^{m-1}n_j)/(\Pi_1^m m_j)$.

Q.E.D.

Remark 2.4.

(i) For the case $\delta = 2$, assuming one has an extension to the walls of Δ (always true if $p = m_0$), the above argument is still valid (see **[W]** for the composition argument, the additivity is a consequence of the additivity up to one suspension of the degree defined in **[I.M.V]** with no action), then

$$deg_K(\tilde{F}) = \Pi_1^m(m_j/m_0) deg_E(f) \quad \text{in } \mathbb{Z}_2.$$

If one of the μ_j's is even then $deg_K(\tilde{F})$ is always 0 and we could not conclude that the Extension degree of the map f is independent of the extension.

(ii) In order to compute the Extension degree of the map f, with f^K S^1-extendable for $K > H$ and $dim_{\mathbb{R}} V^H = dim_{\mathbb{R}} W^H + 1$, it is enough to find a S^1-extension which is non-zero for $z_p = 0$, on the walls of Δ and on B^K. Then $deg_E(f)$ is the degree of this map on the set

$$\left\{0 \le t \le 1, |x_0| \le R, |z| \le R, 0 \le z_p \le R\right\} \cap V^H$$

divided by m_p/m_0, since on each of the m_p/m_0 cells one has the same degree.

2.4. Properties of the Extension degree.

Let H be an isotropy subgroup of S^1 such that $dim_{\mathbb{R}} V^H = dim_{\mathbb{R}} W^H + 1$. Let

$$\Pi = \left\{ [f]_{S^1} \in \Pi^{S^1}_{k+2m}(S^{l+2n}) : f^K : V^K \cap S \longrightarrow W^K \setminus \{0\} \quad \text{is } S^1 - \text{extendable} \quad \text{to} \right.$$
$$\left. B^K \text{ for } \text{all } K > H \right\}.$$

Clearly Π is a subgroup of $\Pi^{S^1}_{k+2m}(S^{l+2n})$.

In fact, using the Borsuk's equivariant extension theorem [**I.M.V, 1.7**], if f and g are S^1- homotopic then f^K has a S^1-extension if and only if g^K has one. Moreover, since the group addition is defined on the invariant part of f, and on the variable t, if f^K and g^K have S^1-extensions then the map $(f+g)^K$ has also a S^1-extension obtained by just *adding* the extensions.

Lemma 2.3. *The assignment*

$$\Pi \overset{deg_E}{\longrightarrow} \mathbb{Z} \qquad (\text{or } \mathbb{Z}_{(\Pi k_j)m_0/p} \text{ if } k = l-1)$$

given by $[f]_{S^1} \longmapsto deg_E(f)$ *is a group homomorphism such that if* $[f]_{S^1} \in ker(deg_E)$ *then* f^H *is* S^1-*extendable. Furthermore, if one has a* S^1-*map,* $g(z_\perp)$, *from* $(V^H)^\perp \setminus \{0\}$ *into* $(W^H)^\perp \setminus \{0\}$ *and if* $F_1(t, x_0, z_H, z_\perp)$, $F_2(t, x_0, z_H, z_\perp)$ *are two maps in* Π *with the same Extension degree, then the maps* $(F_1(t, x_0, z_H, 0), g(z_\perp))$, $(F_2(t, x_0, z_H, 0), g(z_\perp))$ *are* S^1-*homotopic from* S^V *into* $W \setminus \{0\}$.

Proof. Since the Extension degree is unique, the assignment deg_E is well defined. Furthermore, in [**I.M.V, Lemma D.3**] we have proved that deg_K is a morphism (see also Theorem 1.1 (ii)). Thus

$$deg_E(f+g) = deg_K(f+g)/(p/m_0) = (deg_K(f) + deg_K(g))/(p/m_0),$$

giving the additivity of the Extension degree.

Since the maps $f_1 := F_1(t, x_0, z_H, 0)$, $f_2 := F_2(t, x_0, z_H, 0)$ have the same Extension degree, then $deg_E(f_1 - f_2) = 0$, i.e., the map $f_1 - f_2$ is S^1-extendable to $[0,1] \times B_0 \times B^H$.

Then f_1, f_2 are S^1-homotopic on $\partial([0,1] \times B_0 \times B^H)$ (cfr. [**I.M.V**, Remark A.2]) via the homotopy $F_\tau(t, x_0, z_H)$. Extending equivariantly this homotopy (which may have zeros) to $[0,1] \times B_0 \times B^H$ to a S^1-map (with the same notation), then the map $(F_\tau(t, x_0, z_\perp), g(z_\perp))$ gives the desired homotopy.

Q.E.D.

CHAPTER THREE

HOMOTOPY GROUPS OF S^1-MAPS

We shall now begin our study of $\Pi_{k+2m}^{S^1}(S^{l+2n})$ under the dimensional conditions that

(3.1)
$$dim_{\mathbb{R}} V^H \leq dim_{\mathbb{R}} W^H + 1$$

for all isotropy subgroups H of S^1 acting on $V^H = \mathbb{R}^{k+1} \times \mathbb{C}^m$.

Thus in particular, one has for $H = S^1$

$$k + 1 \leq l + 2, \text{i.e.,} k \leq l + 1,$$

and, for $H = \mathbf{Z}_{m_0}$,

$$k + 2m \leq l + 2n + 1,$$

where m_0 is the l.c.d. of $m_1, \ldots, m_m$; note that if W has modes which are not multiples of m_0 one may discard them since all S^1-maps have range in W^H.

If for some H one has the equality in (3.1), then $|k - l|$ is odd. If the strict inequality in (3.1) holds for all H then, from the Corollaries to the Extension Theorem, everything is given by the invariant part of the map. Thus we shall assume that the equality in (3.1) holds at least for one subgroup.

Let then $k = l + 1 - 2p$, $p \geq 0$. For each isotropy subgroup, we shall denote by m_H, n_H the quantities: $k + 1 + 2m_H = dim_{\mathbb{R}} V^H$ and $l + 1 + 2n_H = dim_{\mathbb{R}} W^H$. Thus $m_H \leq n_H + p$ with equality for some H.

We shall restrict our study to the following actions.

Write z as $(z_1, \ldots, z_{m-p}, z_{m-p+1}, \ldots, z_m)$ (if $p = 0$ clearly the last components are omitted), and ξ, in the orthogonal part of $\mathbb{R}^{l+1}$ in W, as $(\xi_1, \ldots, \xi_{m-p}, \xi_{m-p+1}, \ldots, \xi_n)$ ($m \leq n + p$). Assume that for $j = 1, \ldots, m - p$ we have that $n_j = k_j m_j$ and n_j is a multiple of m_r, $r = m - p + 1, \ldots, m$; while for $j \geq m - p + 1$ n_j is arbitrary. (For example, if $p = 0$, $k = l + 1$, one may have that $\mathbb{C}^m = \mathbb{C}^n$ with $k_j = 1$.)

Our peculiar hypothesis implies that if $H = \mathbb{Z}_q$ is an isotropy subgroup and, if no m_r is a multiple of q, for $r = m - p + 1, \ldots, m$, i.e., V^H does not contain the corresponding variables z_r, then $m_H \leq n_H$. Thus, if $p \geq 1$, one has a strict inequality in (3.1), for all $K \geq H$, and an extension to B^H. While, if some m_r is a multiple of q then $n_H \geq m - p \geq m_H - p$ with equality only if $m = m_H$ that is only if $q = m_0 = (m_1 : \ldots : m_m)$, H is the smallest subgroup and $n = m - p$. Thus, if $k = l + 1 - 2p, p \geq 0, m \leq n + p$, the hypothesis on the actions implies the dimension inequalities (3.1). Furthermore, if $p \geq 1$ and $m - p < n$, then from Corollary 2.1, $\Pi_{k+2m}^{S^1}(S^{2+2n}) = 0$, that is any S^1-map has an extension to B^{k+1+2m}, while if $p \geq 1$ and $m - p = n$, then one has S^1−extensions to B^k for any isotropy subgroup except for $\mathbb{Z}_{m_0}$, with m_0 the l.c.d. of all the $m'_j s$. In this last case, $\Pi = \Pi_{k+2m}^{S^1}(S^{l+2n})$ and, from Lemma 2.3, deg_E will characterize Π. In the following result we shall compute the range of this morphism.

3.1. Trivial invariant part, the case $p \geq 1$.

Theorem 3.1. If $k = l + 1 - 2p$, $p \geq 1$, $n_j = k_j m_j$ for $j = 1, \ldots, m - p = n$, n_j multiple of m_r, $r = n + 1, \ldots, m$, then

$$\Pi_{k+2m}^{S^1}(S^{l+2n}) = \begin{cases} \mathbb{Z}, & p > 1 \\ \mathbb{Z}_{(\Pi k_j) m_0 / m_m}, & p = 1 \ (0, \text{if } n = 0, m = 1). \end{cases}$$

Proof. a) If $p = 1$, one knows from Lemma 2.2 that two extensions may differ by a multiple of $(\Pi k_j) m_0 / m_m$, however the following explicit construction will be needed. Let $(\tilde{f}_0, \tilde{\varphi}_0)$ be any extension of $(1, 0)$ to $[0, 1] \times B_0$ with norm 1 and of degree d. Consider the map

$$f_d(t, x_0, z) = \left(\frac{R - |z|}{R} (\tilde{f}_0 - 1, \tilde{\varphi}_0, \dots, (z_j^{n_j/m_j} - \varepsilon_j z_m^{n_j/m_m}) t (1 - t)(R_0 - |x_0|), \dots \right)$$
$$+ (1, 0, 0)$$

where j runs from 1 to $n = m - 1$ (as in the proof of Theorem 4.6 of [**I.M.V**, Appendix D]). The above map is equivariant and is $(1, 0, 0)$ on ∂B. Thus $[f_d] = 0$. Now f_d has its zeros at $\tilde{\varphi}_0 = 0$ (hence $\tilde{f}_0$ is either $+1$ or -1), $z_j^{n_j/m_j} = \varepsilon_j z_m^{n_j/m_m}$ and $|z| = R/2$ ($|\varepsilon_j|$ so small that the zeros are in B). Thus $|z_m| \neq 0$ and $|z_j| \neq 0$ that is, this map is non-zero on any isotropy subspace V^K, $K > \mathbb{Z}_{m_0}$ (m_0 the l.c.d. of $m_1, \dots, m_m$). For $|\varepsilon_j|$ small, $|z| = max |z_j| = |z_m|$; thus for $0 \leq z_m \leq R$, one has $\Pi_1^n n_j/m_j$ zeros of the equation $z_j^{n_j/m_j} = \varepsilon_j (\frac{R}{2})^{n_j/m_m}$ and one may choose the phases of ε_j such that none of these zeros are on the faces of Δ. By replacing z_j by $Z_j^{m_j/m_0}$, $j = 1, \dots, m$, one may compute $deg_K(\Theta^* f_d)$, by calculating the degree for $0 \leq Z_m \leq R$. By deforming the terms $Z_m^{n_j/m_0}$ to 0, one obtains the map defined in [**I.M.V**, Appendix D] with degree equal to $d\Pi_1^n(n_j/m_0)$. Hence, according to Lemma 2.2,

$$deg_E(f_d) = d\Pi_1^n(n_j/m_0)\Pi_1^m(m_0/m_j) = d(\Pi_1^n k_j)(m_0/m_m)$$

(the above number is an integer since if m_m/m_0 would not divide any of the k_j's, then from $n_j/m_0 = k_j(m_j/m_0) = \tilde{k}_j(m_m/m_0)$, we have that m_m/m_0 would have to divide all m_j/m_0 and m_0 would not be the l.c.d. of $m_1, \dots, m_m$).

This construction implies that one may always add to any S^1-map f a map which is S^1-trivial but with an Extension degree equal to any multiple of $(\Pi k_j)m_0/m_m$. Thus the Extension degree will be defined modulo $(\Pi k_j)m_0/m_m$.

Take now a positive integer d and let l_j be the l.c.d. of $m_m, m_1, \dots, m_j$. Thus from Darboux theorem, there are integers $\alpha_{m,j}, \alpha_{1,j}, \dots, \alpha_{j,j}$ such that

$$\alpha_{m,j} m_m + \sum_{q=1}^{j} \alpha_{q,j} m_q = l_j.$$

Consider then the S^1-map $F_d(t, x_0, z) \equiv$

$$(1 - \frac{2}{R}|z_m|, |z_m|(2t - 1), |z_m|x_0, z_m^{n_1/m_m}[z_m^{dm_1/l_1} z_1^{dm_m/l_1} - \varepsilon_1], \dots,$$
$$z_m^{n_j/m_m}\left[(\bar{z}_m^{\alpha_{m,j}-1} \Pi_{q=1}^{j-1} \bar{z}_q^{\alpha_{q,j}-1})^{m_j/l_j} z_j^{l_j-1/l_j} - \varepsilon_j \right]_{j=2,\dots,n})$$

where z^α for negative α means $\bar{z}^{|\alpha|}$. Note that the terms in the square brackets are invariant.

The zeros of the map F_d are for $|z_m| = R/2$, $t = 1/2$, $x_0 = 0$, $|z_j|^{l_{j-1}/l_j} = |\varepsilon_j| |\eta_j|$, where $|\eta_j|$ depends on $|z_m|, |\varepsilon_1|, \ldots, |\varepsilon_{j-1}|$ and is non-zero if these arguments are non-zero. Thus, taking $\varepsilon_j \neq 0$ the zeros are in the set $(V \setminus \bigcup_{K > \mathbb{Z}_{m_0}} V^K) \setminus \{0\}$. One may choose $|\varepsilon_j|$ such that the zeros are in B. Then for $z_m = R/2$ one has $(dm_m/l_1)(l_1/l_2) \ldots (l_{n-1}/l_n) = dm_m/m_0$ distinct zeros such that the norm of each component is independent of the zero. Since there is a finite number of zeros, one may choose inductively the phases of ε_j such that none of them is on the walls of Δ. In this case the Extension degree will be the Brouwer degree of $F_d(t, x_0, z)$ on the set $[0,1] \times B_0 \times \{B \cap \{0 \le z_m \le R\}\}$ divided by the number of cells, that is by m_m/m_0.

By the product formula for the degree one gets

$$(-1)^{k+2} deg \left(\left(z_1^{dm_m/l_1} - \varepsilon_1', \ldots, (\Pi_{q=1}^{j-1} \bar{z}_q^{\alpha_{q,j-1}})^{m_j/l_j} z_j^{l_{j-1}/l_j} - \varepsilon_j', \ldots \right) \right) (m_0/m_m).$$

Let $(z_1^0, \ldots, z_n^0)$ be one of the dm_m/m_0 zeros of the above map. In order to compute the local index at this point, one may replace, in $\Pi_{q=1}^{j-1} \bar{z}_q^{\alpha_{q,j-1}}$, each z_q by $tz_q + (1-t)z_q^0$ without moving the zero. Thus the local index is the index of the map

$$\left(z_1^{dm_m/l_1} - \varepsilon_1', \delta_1 z_2^{l_1/l_2} - \varepsilon_2', \ldots, \delta_n z_n^{l_{n-1}/m_0} - \varepsilon_n' \right).$$

This is an analytic map, hence with index 1 (there is just one zero in the neighborhood of $(z_1^0, \ldots, z_n^0)$). Thus the degree of the map with arguments $(z_1, \ldots, z_n)$ is dm_m/m_0 and the Extension degree is $(-1)^k d$.

For negative d, one constructs a similar map with $2t - 1$ replaced by $1 - 2t$ and d replaced by $|d|$. One gets an Extension degree equal to $(-1)^{k-1}|d| = (-1)^k d$. Thus one obtains all Extension degrees.

If $n = 0$, then $m = 1$ and one may always extend, see Theorem 1.1 (ii) (c).

b) If $p > 1$, then the Extension degree is well defined.

For $n \ge 0$ and $d > 0$ consider the map $F_d(t, x_0, z) =$

$$\left(1 - \frac{2}{R}|z_m|, |z_m|(2t-1), |z_m|x_0, \left(\Pi_{q=1}^n \bar{z}_q^{\alpha_{q,n}} \bar{z}_m^{\alpha_{m,n}}\right)^{dm_{n+1}/l_{n+1}} z_{n+1}^{dl_n/l_{n+1}} - \varepsilon_{n+1},\right.$$

$$\left\{\left(\Pi_{q=1}^{m+j} \bar{z}_q^{\alpha_{q,n+j}} \bar{z}_m^{\alpha_{m,n+j}}\right)^{m_{n+j+1}/l_{n+j+1}} z_{n+j+1}^{l_{n+j}/l_{n+j+1}} - \varepsilon_{n+j+1}\right\}_{j=1,\dots,p-2},$$

$$\bar{z}_m^{n_1/m_m}\left(\bar{z}_m^{m_1/l_1} z_1^{m_m/l_1} - \varepsilon_1\right),$$

$$\left.\left\{\bar{z}_m^{n_j/m_m}\left[\left(\Pi_{q=1}^{j-1} \bar{z}_q^{\alpha_{q,j-1}} \bar{z}_m^{\alpha_{m,j-1}}\right)^{m_j/l_j} z_j^{l_{j-1}/l_j} - \varepsilon_j\right]\right\}_{j=2,\dots,n}\right)$$

where l_j and ε_j are as before.

Clearly the arguments given for the case $p = 1$ are still valid and one will get an Extension degree equal to $(-1)^k d$ (including the case when $n = 0$, reason for which the d term is on the invariant part). For d negative, changing $2t - 1$ to $1 - 2t$ one gets all the degrees.   Q.E.D.

Remark 3.1. If $k = l + 1 - 2p$ with $p \geq 1$, then there is no practical gain in these results with respect to the invariants in $\Pi_{k+2m,\bullet}^{S^1}(S^{l+2n})$ since the two degrees are related by $\deg_K(\Theta^* f) = (\Pi_1^m m_j / m_0) \deg_E(f)$. On the other hand if one forgets the action, then one has two elements in $\Pi_{k+2m}(S^{l+2n}) = \mathbb{Z}_2$, one of which is equal to $d'(\sum n_j / m_0 + m + d')\Sigma^{l+2n}\eta$, where η is the Hopf map, if $\deg_K(\Theta^* f) = d'$ (cfr. **[I.M.V, Corollary 4.7]**). Clearly the following diagram is commutative:

$$
\begin{array}{ccc}
\Pi_{k+2m}^{S^1}(S^{l+2n}) & \xrightarrow{\ \Theta^*\ } & \Pi_{k+2m,\bullet}^{S^1}(S^{l+2n}) \\[4pt]
\downarrow \chi_* & & \downarrow \chi_* \\[4pt]
\Pi_{k+2m}(S^{l+2n}) & \xrightarrow{\ \Theta^*\ } & \Pi_{k+2m}(S^{l+2m})
\end{array}
$$

Since $\chi_* \Theta^* f = 0$ if m_j / m_0 is even for some j, there is still the question to know if in this case $\chi_*(f) \neq 0$. However the computation of this last invariant is not easy and, since this point is not crucial for this paper, we shall compute it only in a particular case.

As an example take the case when $m_m = m_0$. Then, in the proof of the previous theorem, one has $l_j = m_0$, $\alpha_{q,j} = 0$, $q \neq m$, $\alpha_{m,j} = 1$ for all j' s. The map is then

$$\left(1 - \frac{2}{R}|z_m|, |z_m|(2t-1), |z_m|x_0, \bar{z}_m^{dm_{n+1}/m_0} z_{n+1}^d - \varepsilon_{n+1},\right.$$

$$\left.\{\bar{z}_m^{m_{n+j}/m_0} z_{n+j} - \varepsilon_{n+j}\}_{j=2,\dots,p-2}\{\bar{z}_m^{n_j/m_0}(\bar{z}_m^{m_j/m_0} z_j - \varepsilon_j)\}_{j=1,\dots,n}\right)$$

(if $p = 1$, $n \geq 1$, then the term $z_m^{dm_{n+1}/m_0} z_{n+1}^d$ is transfered to $z_m^{dm_1/m_0} z_1^{d_0}$). By deforming ε_j to 0 and performing rotations (as in the proof of Corollary 4.7 in [**I.M.V**, Appendix D]) one gets the class of the map

$$(1 - \frac{2}{R}|z_m|, 2t - 1, x_0, z_m^\alpha z_m^\beta z_{n+1}^d, z_{n+2}, \ldots, z_{n-1}, z_1, \ldots, z_n),$$

where $\alpha = \sum_1^n n_j/m_0$ and $\beta = \left(\sum_1^{m-1} m_j/m_0 \right) + (d-1)m_{n+1}/m_0$.

(If $p = 1$ and $m = n + 1$, one has the map $z_m^\gamma z_1^d$ with $\gamma = \sum_1^n (n_j - m_j)/m_0$ $-(d-1)m_1/m_0$.)

The class of this map is

$$d \left(\sum_1^n n_j/m_0 + \sum_1^{m-1} (m_j/m_0) + (d-1)m_{n+1}/m_0 \right) \Sigma^{l+2n} \eta.$$

If d is even, this element is 0, while if d is odd the element will be non-zero if and only if $\sum_1^n (n_j/m_0) + \sum_1^{m-1} (m_j/m_0)$ is odd (for example if $n_j = m_j$, $j = 1, \ldots, n$, the element is non-zero if and only if $\sum_{n+1}^{m-1} m_j/m_0$ is odd).

Thus if d is odd but one m_j/m_0 is even, then $\chi_*(\Theta^* f) = 0$ but $\chi_*(f) \neq 0$ if the two sums have different parity. If d is odd and all m_j/m_0 are odd then $m + d'$ and $\sum_1^{m-1} m_j/m_0$ have both the parity of $m - 1$ and thus one does not lose information.

Note that if $p = 1$, then the map of degree $d = \Pi_1^n n_j/m_0$ generates an element which is 0. Indeed if the product is odd, then each $n_j/m_0 = k_j m_j/m_0$ is odd and thus each m_j/m_0 is odd, and their sum, from 1 to n, is even.

3.2. Nontrivial invariant part, the case $p = 0$.

We shall now study the case $k = l + 1$, ($p = 0$) under the assumptions that $n_j = k_j m_j$, $j = 1, \ldots m$. Then $m_H \leq n_H$ for all H, in particular $m \leq n$.

We are now in the position to prove the following result.

Theorem 3.2. *Suppose that* $n_j = k_j m_j$, $k = l+1$, *then*

$$\Pi^{S^1}_{k+2m}(S^{l+2n}) \cong \Pi_k(S^l) \times \mathbb{Z} \times \ldots \times \mathbb{Z}$$

with one $\mathbb{Z}$ *for each isotropy subgroup* H *such that* $m_H = n_H$.

Proof. Let $F(t, x_0, z) = (f_0, \Phi_0, \Phi)(t, x_0, z)$ be a representative of an element, $[F]_{S^1}$, in $\Pi^{S^1}_{k+2m}(S^{l+2n})$, (with $f_0(1,0) > 0$ in the case $l = 0$), where one may assume that $F(1, x_0, z) = F(0, x_0, z) = (1, 0, 0)$.

0) Invariant part.

One has that:

$$P_*[F]_{S^1} = [(f_0, \Phi_0)(t, x_0, 0)] = d_0 \Sigma^{l-2} \eta \qquad \text{if } l \geq 2$$
$$= 0 \qquad \text{if } l = 0, 1$$

where η is the Hopf map, represented in the coordinates (t, x_1, x_2, x_3) by

$$\big((2t-1)^2 + x_1^2 - x_2^2 - x_3^2, ((2t-1) - ix_1)(x_2 + ix_3)\big)$$

(the map $d_0\eta$ is obtained by replacing $x_2 + ix_3$ with $(x_2 + ix_3)^{d_0}$).

Let F_0 be the following S^1-map: $\big((\Sigma^{l-2}\eta)(t, x_0), z_1^{k_1}, \ldots, z_m^{k_m}, 0, \ldots, 0\big)$ and for $x_0 = (t, x_1, x_2, x_3, \tilde{x}_0)$, let $d\tilde{F}_0$ be the map

$$d\tilde{F}_0 = \big((2t-1)^2, 0, \ldots, 0\big) + t(1-t)(x_1^2 - x_2^2 - x_3^2, ((2t-1) - ix_1)(x_2 + ix_3)^d,$$
$$\tilde{x}_0, z_1^{k_1}, \ldots, z_m^{k_m}, 0, \ldots, 0\big).$$

Since P_* is a morphism (cfr. **[I.M.V**, Lemma 4.3]), the map

$$F_1(t, x_0, z) \equiv \begin{cases} F(2t, x_0, z) & 0 \leq t \leq 1/2 \\ d_0\tilde{F}_0(2 - 2t, x_0, z) & 1/2 \leq t \leq 1 \end{cases}$$

is such that $[F_1]_{S^1} = [F]_{S^1} - d_0[F_0]_{S^1}$ and, since $P_*[F_1]_{S^1} = 0$, F_1 has a S^1-extension to $I \times B_0 \times \{z = 0\}$, (note that $d_0 = 0$ if $l = 0$ or 1).

1) First isotropy subgroup.

Let $m_0 \equiv \max m_j$ and $H_1 = \mathbb{Z}_{m_0}$ be the corresponding isotropy subgroup with the respective dimensions m_{H_1} and n_{H_1}.

Write $z = (\tilde{z}, \hat{z}) \equiv (z_1, \ldots, z_r, z_{r+1}, \ldots, z_m)$, with $\tilde{z}$ corresponding to the vectors in V^{H_1} and $r = m_{H_1}$.

If $m_{H_1} < n_{H_1}$, then F_1 S^1-extends to $I \times B_0 \times B^{H_1}$. Assume that $m_{H_1} = n_{H_1}$, then F_1 has a unique extension degree d_1.

a) For $l \geq 1$, write $x_0 = (x, y, \tilde{x}_0)$, with $\tilde{x}_0$ in $\mathbb{R}^{l-1}$ and set $\lambda = x + iy$. Define then

$$(-1)^k dF_{H_1} \equiv (1,0,0) + 4t(1-t)(-2|z_1|/R, 2t-1, \tilde{x}_0, \lambda^d z_1^{k_1}, z_1^{k_2}\bar{z}_1 z_2, \ldots,$$
$$z_1^{k_r}\bar{z}_1 z_r, z_{r+1}^{k_{r+1}}, \ldots, z_m^{k_m}, 0, \ldots, 0).$$

(Again λ^{-1} means $\bar{\lambda}$).

Clearly dF_{H_1} is $(1,0,0)$ for $t = 0$ or 1, and is zero only if $t = 1/2, \tilde{x}_0 = 0, |z_1| = R/2, \lambda = 0, z_j = 0$ for $j = 2, \ldots, m$. Thus the extension degree on V^{H_1} is just the Brouwer degree of its restriction to the set $I \times B_0 \times \{0 \leq z_1 \leq R, |z_j| \leq R, j = 2, \ldots, r, z_{r+j} = 0\}$.

Deforming $4t(1-t)$ to 1, it is easy to see that the extension degree is $(-1)^k d$, the sign coming from the orientation.

Then, since the extension degree is a morphism (Lemma 2.3) one has that $[dF_{H_1}]_{S^1} = d[F_{H_1}]_{S^1}$ (the map $g(z_\perp)$ is $(z_{r+1}^{k_{r+1}}, \ldots, z_m^{k_m}, 0, \ldots, 0)$).

b) For $l = 0$ and $k = 1$, consider the map:

$$dF_{H_1} \equiv (1,0,0) + 8t(1-t)\{(x_0/R + (2|z_1| - R)^2/R^2,$$
$$(2t-1-i(2|z_1|-R))^d z_1^{k_1}, z_1^{n_2/m_1}\bar{z}_1^{m_2/m_1} z_2, \ldots,$$
$$z_1^{n_r/m_1}\bar{z}_1^{m_r/m_1} z_r, z_{r+1}^{k_{r+1}}, \ldots, z_m^{k_m}, 0, \ldots, 0)\}.$$

(Note that $m_1 = \ldots = m_r = m_0$ for this isotropy subgroup. The above notation is given for later use).

Since $x_0/R \geq -1$, the zeros of this map in $I \times B_0 \times B$ are for $|z_1| = R/2, t = 1/2, z_j = 0$ for $j \geq 2$, and $x_0 = -R/2$ (recall that $|z_j| \leq R$).

One may deform the factor $8t(1-t)$ to 2 and, for $0 \le z_1 \le R$, the terms $z_1^{k_1}, z_1^{(n_j-m_j)/m_1}$ to 1. Then the extension degree is the degree of the map:

$$(t, x_0, z_1) \to (1 + 2x_0/R, (2t - 1 - i(2z_1 - R))^d)$$

which is d.

Thus $[F_2]_{S^1} \equiv [F]_{S^1} - d_0[F_0]_{S^1} - d[F_1]_{S^1}$ has an extension to B^{H_1}.

s) The s'th isotropy subgroup.

Let H_s be an arbitrary isotropy subgroup, with $H_s = \mathbb{Z}_{m_s}$.

Suppose that

$$[F_s]_{S^1} \equiv [F]_{S^1} - d_0[F_0]_{S^1} - d_1[F_1]_{S^1} \ldots - d_{s-1}[F_{s-1}]_{S^1}$$

has been constructed in such a way that F_s has non-zero S^1-extensions to $I \times B_0 \times B^K$ for all K's, $K > H_s$ and for the isotropy subgroups preceding H_s in the decreasing order of m_s. As before if $m_{H_s} < n_{H_s}$, then F_s has an extension to B^{H_s}.

If $m_{H_s} = n_{H_s} \equiv r$, write $z = (\bar{z}, \hat{z}) = (z_1, \ldots, z_r, z_{r+1}, \ldots, z_m)$, with $m_1 \le m_2 \le \ldots \le m_r$ are the only multiples of m_s.

Then F_s has a unique extension degree d_s, such that F_s extends to B^{H_s} if and only if $d_s = 0$. Furthermore the extension degree is a morphism for the maps which have extensions to $B^K, K > H_s$.

Since m_s is the l.c.d. of $(m_1, \ldots, m_r)$, there are two possible cases: $m_1 = m_s$ or $m_1 = qm_s, q > 1$.

a) Case $m_1 = m_s$.

α) If $l \ge 1$, let $x_0 = (x, y, \tilde{x}_0), \tilde{x}_0 \in \mathbb{R}^{l-1}, x + iy = \lambda$, and define the map:

$$(-1)^k dF_{H_s} = (1, 0, 0) + 4t(1 - t)\{(-2|z_1|/R, 2t - 1, \tilde{x}_0, \lambda^d z_1^{k_1},$$
$$z_1^{n_2/m_1} \bar{z}_1^{m_2/m_1} z_2, \ldots, z_1^{n_r/m_1} \bar{z}_1^{m_r/m_1} z_r, z_{r+1}^{k_{r+1}}, \ldots, z_m^{k_m}, 0, \ldots, 0)\},$$

$(\lambda^{-1}$ means $\bar{\lambda})$.

Note that we have ordered the components of $\tilde{z}$ and $\hat{z}$, as well as the components of the range, in this special way just for the sake of notational clarity. Since $z_j, j \geq 2$, are complex, then the extension degree is unchanged when one interchanges the positions of these components.

It is easy to see, as before, that the extension degree of dF_{H_s} is d.

$\beta)$ If $l = 0$, consider the map dF_{H_s} written as dF_{H_1} in the case $H = H_1$.

In both cases $[dF_{H_s}]_{S^1} = d[F_{H_s}]_{S^1}$.

b) Case $m_1 = qm_s, q > 1$.

Let l_j be the l.c.d. of $(m_1, \ldots, m_j), j = 1, \ldots, r$. Thus $l_1 = m_1, l_r = m_s$. Let $\alpha_{q,j}$ be such that $\sum_1^j \alpha_{q,j} m_q = l_j$.

$\gamma)$ For $l \geq 1$, consider the map:

$$(-1)^k dF_{H_s} = (1, 0, 0) + \varphi(t)\{(-2|z_1|/R, 2|z_2| \ldots |z_r|t - (R/2)^{r-1},$$
$$\tilde{x}_0, \lambda^d z_1^{k_1}, z_2^{k_2}(\varphi(t)^{m_2/l_2} \bar{z}_1^{m_2/l_2} z_2^{m_1/l_2} - \varepsilon_2), \ldots,$$
$$z_j^{k_j}([\varphi(t)^{\alpha_{1,j}-1}\Pi_{q=1}^{j-1}\bar{z}_q^{\alpha_{q,j}-1}]m_j/l_j z_j^{l_{j-1}/l_j} - \varepsilon_j)_{j=3,\ldots,r},$$
$$z_{j+r}^{k_{j+r}}, j = 1, \ldots, m - r, 0, \ldots, 0)\},$$

where $\varphi(t) = 4t(1 - t)$ and ε_j will be given below.

The zeros of this map are for $\tilde{x}_0 = 0, \lambda \neq 0, \varphi(t)|z_1| = R/2$ (thus one cannot have any of the $z_j = 0, j = 1, \ldots, r$, considering the second component), $|z_2| = |\varepsilon_2|^{l_2/m_1}/(R/2)^{m_2/m_1} \equiv |\eta_2|, |z_j| = |\eta_j|, j = 2, \ldots, r, z_{r+j} = 0, j = 1, \ldots, m - r, 2|\eta_2| \ldots |\eta_r| = (R/2)^{r-1}$.

Note that $|\eta_j|$ doesn't depend on t. Choose $|\varepsilon_2|, \ldots, |\varepsilon_r|$ such that $|\eta_2| = \ldots = |\eta_r| = R/2$. Thus for a zero one has $t = 1/2$ and $|z_1| = R/2$.

Choose the phases of $\varepsilon_2, \ldots, \varepsilon_r$ such that, for $z_1 = R/2$, the $m_1/m_s = q$ different zeros are not on the walls of the fundamental cell $\triangle$.

As in Remark 2.4, one may compute directly the extension degree of this map, by dividing by q the degree of the map on the set $I \times B_0 \times B^H \cap \{0 \leq z_1 \leq R\}$.

In order to compute this degree, one may deform $\varphi(t)$ to 1 via $\varphi(\tau t + (1-\tau)/2)$, recalling that $|\eta_j|$ does not depend on t. Deforming next z_1 to $R/2$ in all components but the first and $|z_j|$ to $R/2$ in the second component, one has to compute the product of degrees of the maps:

$$(t, \lambda, \tilde{x}_0, z_1) \to (1 - 2z_1/R, 2t - 1, \tilde{x}_0, \lambda^d)$$

and

$$(z_2, \ldots, z_r) \to (\ldots, z_j^{k_j} \left((\Pi_{q=1}^{j-1} \bar{z}_q^{\alpha_{q,j-1}})^{m_j/l_j} z_j^{l_{j-1}/l_j} - \varepsilon_j \right), \ldots)$$

where $z_1 = \eta_1 = R/2$. Clearly the first map has degree $(-1)^{l-1}d$. The second map has degree q. In fact, one may compute it in the following way: in a neighborhood of the zero $(\eta_2, \ldots, \eta_r)$, one may deform this map to $(\ldots, \eta_j^{kj}(\beta_j z_j^{l_{j-1}/l_j} - \varepsilon_j), \ldots)$, where $\beta_j = \left(\Pi_1^{j-1} \bar{\eta}_j^{\alpha_{q,j-1}} \right)^{m_j/l_j} = \varepsilon_j \eta_j^{-l_{j-1}/l_j}$. Then one deforms to the map $(\ldots, (z_j/\eta_j)^{l_{j-1}/l_j} - 1, \ldots)$ with index 1 at $(\eta_2, \ldots, \eta_r)$.

Then the extension degree of $(-1)^k dF_{H_s}$ is $(-1)^k d$. Note that one may use this map for the case $m_1 = m_s$, by taking $\alpha_{1,j-1} = 1, \alpha_{q,j-1} = 0$ for $q > 1$ and $l_{j-1} = l_j = m_1$.

δ) If $l = 0$, consider the map:

$$dF_{H_s} \equiv (1, 0, 0) + 8t(1-t)\{(x_0/R + \sum_{j=1}^{r} (2|z_j| - R)^2/R^2,$$

$$((2t-1) - i(2|z_1| - R))^d z_1^{k_1}, z_j^{k_j}((\Pi_{q=1}^{j-1} \bar{z}_q^{\alpha_{q,j-1}})^{m_j/l_j} z_j^{l_{j-1}/l_j} - \varepsilon_j)_{j=2,\ldots,r},$$

$$z_{j+r}^{k_{j+r}}, 0, \ldots, 0)\}.$$

Since $x_0 \geq -R$, none of the z_j's may be 0 for a zero of the map. For such a zero, one has $|z_1| = R/2, t = 1/2$ and, choosing $|\varepsilon_j|$ appropriately as before, $|z_j| = |\eta_j| = R/2$ and $x_0 = -R/2$.

For $0 \leq z_1 \leq R$, the degree of this map is d times the degree of the map:

$$(x_0, z_2, \ldots, z_r) \to (1 + 2x_0/R + 2\sum_2^r(2|z_j| - R)^2/R^2, z_j^{k_j}\left((\Pi_{q=1}^{j-1}\bar{z}_q^{\alpha_{q,j-1}})^{m_j/l_j}z_j^{l_{j-1}/l_j} - \varepsilon_j\right),$$

where z_1 is replaced by $R/2$.

At each of the $m_1/m_s = q$ zeros of this map, of the form $(-R/2, \eta_2, \ldots, \eta_r)$, one has, as before, an index equal to 1 and thus the extension degree of dF_{H_s} is d.

In both cases $[dF_{H_s}]_{S^1} = d[F_{H_s}]_{S^1}$.

c) The map $[F_{s+1}]_{S^1} \equiv [F]_{S^1} - \sum_0^s d_j[F_{H_j}]_{S^1}$ has a S^1-extension to $I \times B_0 \times B^k$ for all $K \geq H_s$ and also for all K 's corresponding to $\mathbb{Z}_{m_0}$ with $m_0 \geq m_s$.

Thus, once one has exhausted all the isotropy subgroups (the last is for m_0 the l.c.d. of $m_1, \ldots, m_m$), the last map after say s steps, where s is the number of isotropy subgroups, is extendable to $I \times B_0 \times B$ that is, it is S^1-trivial. Thus

$$[F]_{S^1} = \sum_0^s d_j[F_{H_j}]_{S^1},$$

where $d_j = 0$ whenever $m_{H_j} < n_{H_j}$.

From our step by step construction and from the facts that at each step the extension degree is unique and that $[dF_{H_j}]_{S^1} = d[F_{H_j}]_{S^1}$, it is easy to see that the assignment $[F]_{s^1} \to (d_0, d_1, \ldots, d_s)$ is well defined, i.e., independent of the representative F, and that it is a morphism, which is one to one and, by construction, onto $\Pi_k(S^l) \times \mathbb{Z} \times \ldots \times \mathbb{Z}$.

Q.E.D.

3.3. Behavior under suspension.

As in [**I.M.V**] we shall need to see what happens to the group $\Pi_{k+2m}^{S^1}(S^{l+2n})$ when one adds a variable. This suspension will be important when one works with infinite dimensional problems.

There are two possible ways to suspend a map: with an invariant real variable x or with a complex variable z with action $e^{iq\varphi}$ and giving z^r in the range (usually $r = 1$).

We shall assume that we are in a position of applying our previous two results; hence for the case $k = l + 1 - 2p$ with $p \geq 1$, we shall consider, for the suspension z^r, only the case when

$$qr \text{ is a multiple of } m_j \text{ for } j = m - p + 1, \ldots, m.$$

In this case the hypothesis on the actions is preserved. Let

$$\Sigma_0 : \Pi_{k+2m}^{S^1}(S^{l+2n}) \to \Pi_{k+1+2m}^{S^1}(S^{l+1+2n})$$

be the suspension of a map F by an invariant real variable x and let

$$\Sigma_r : \Pi_{k+2m}^{S^1}(S^{l+2n}) \to \Pi_{k+2(m+1)}^{S^1}(S^{l+2(n+1)})$$

be the suspension, under the above hypothesis, by the addition of the assignment $z \to z^r$.

Clearly both maps are morphisms. Furthermore Σ_0 doesn't alter the isotropy subgroups of V and of $V \times \mathbb{R}$.

On the other hand this is not true anymore for V and $V \times \mathbb{C}$. In fact if $H = \mathbb{Z}_{m_0}$, with $m_0 = |H|$ the order of H, is an isotropy subgroup for V, then H will remain an isotropy subgroup for $V \times \mathbb{C}$ but $(V \times \mathbb{C})^H = V^H$, if m_0 doesn't divide q, while $(V \times \mathbb{C})^H = V^H \times \mathbb{C}$ if m_0 divides q. Furthermore $V \times \mathbb{C}$ will have also as isotropy subgroups (if not included in those of V) $\mathbb{Z}_q$ and $\mathbb{Z}_{m_0}$ with m_0 the l.c.d. of q and $|H_j|$, for some isotropy subgroup H_j for V.

Let then $H_j, j = 0, \ldots, s$, be the isotropy subgroups for $V(H_0 = S^1)$. Divide the set of indices in the following subsets:

$$I_0 = \{0, j \text{ such that } |H_j| \text{ doesn't divide } q \},$$
$$I_q = \{j \text{ such that } |H_j| \text{ divides } q\}$$
$$I_{qr} = \{j \text{ such that } |H_j| \text{ divides } qr \text{ but not } q\}.$$

For each H_j, let $\bar{H}_j$ be the group with order equal to the l.c.d. of q and $|H_j|$.

Note that one may have $\bar{H}_j = H_{j_0}$ (that is the l.c.d. $(q : |H_j|) = |H_{j_0}|$) for some subgroup for V or one may also have $\bar{H}_j = \bar{H}_{j_0}$ for two different subgroups H_j and H_{j_0}.

Thus the additional subgroups for $V \times \mathbb{C}$, H_{j+s}, will be one of these $\bar{H}_j$ or $\mathbb{Z}_q$ if q doesn't divide any m_j. Here $j + s$ runs from $1 + s$ to $t + s$, with $0 \le t \le s + 1$.

If $k = l + 1$, then any F in $\Pi_{k+2m}^{S^1}(S^{l+2n})$ can be written as:

$$[F]_{S^1} = \sum_0^s d_j [F_{H_j}]_{S^1}, d_j = d_j(F),$$

with $d_j(F) = 0$, if $m_{H_j} < n_{H_j}$, and F_{H_j} is the generator corresponding to H_j and given in the proof of Theorem 3.2.

Similarly any element $\tilde{F}$ of $\Pi_{k+2(m+1)}^{S^1}(S^{l+2(n+1)})$ can be written as:

$$[\tilde{F}]_{S^1} = \sum_0^{s+t} \tilde{d}_j [\tilde{F}_{H_j}]_{S^1}, \tilde{d}_j = \tilde{d}_j(\tilde{F}),$$

where $\tilde{F}_{H_j}$ are the corresponding generators.

For each $j = 1, \ldots, s$, let $\tilde{F}_{\bar{H}_j}$ be the generator corresponding to $\bar{H}_j$.

One has $V^{H_j} \subset V^{\bar{H}_j}$; in case of a strict inequality denote by $q_j = \Pi k_s$, where the product is over all indices s such that m_s divides $|\bar{H}_j|$ but doesn't divide $|H_j|$. (In order to have a unique notation, for Σ_0 one has $t = 0$).

We are now ready for the following **Suspension Theorem:**

Theorem 3.3.

1) Σ_0 is an isomorphism except for the cases $k = l + 1$, with $l = 1$ or 2. In these cases, if $l = 1$, then $d_0(F) = \tilde{d}_0(\Sigma_0 F) = 0$ and if $l = 2$, then $\tilde{d}_0(\Sigma_0 F) \equiv d_0(F)$ modulo 2, while $\tilde{d}_j(\Sigma_0 F) = d_j(F)$ for $j = 1, \ldots, s$.

2) If $k \le l - 1$, let $l_0 = (m_1 : \ldots : m_m)$ and $m_0 = (l_0 : q)$ be the l.c.d.'s of these numbers. Then $m_0 r / l_0$ is an integer and

$$deg_E (\Sigma_r F) = (m_0 r / l_0) deg_E(F).$$

Thus, Σ_r is always one to one and Σ_r is onto only if $m_0 r = l_0$, with $n > 0$ if $k = l - 1$. If $r = 1$, then $m_0 = l_0$ and Σ_1 is an isomorphism, ($n > 0$ if $k = l - 1$).

3) If $k = l + 1$, then with the above notation:

$$[\Sigma_r F]_{S^1} = \sum_{I_0} d_j [\tilde{F}_{H_j}]_{S^1} + r \sum_{I_q} d_j [\tilde{F}_{H_j}]_{S^1} +$$

$$r \sum_{I_{qr}} d_j q_j |\bar{H}_j|/|H_j| [\tilde{F}_{\bar{H}_j}]_{S^1}.$$

(recalling that the generators of the last sum may not be independent and may also be generators of the first sums).

In particular if $r = 1$, then $I_{qr} = \emptyset$ and Σ_1 is a one to one map. Σ_1 is onto only if $q = |H_j|$ for some j.

Proof. 1) If $p \geq 1$ then, by suspending by x, one changes x_0 to (x_0, x) without affecting the result (see Theorem 3.1). If $p = 0$ and $l \geq 1$, then clearly $\tilde{F}_{H_j} = \Sigma_0 F_{H_j}$ with the exception of $S^1 = H_0$ when $l = 1$ (in this case $[F_{H_0}]_{S^1} = 0$); thus $\tilde{d}_j(\Sigma_0 F) = d_j(F)$ for $j \geq 1$ (the suspended degree is unchanged), while, for d_0, one has the usual results for $\Pi_{l+1}(S^l)$. If $p = 0$ and $l = 0$, $[F_{H_0}]_{S^1} = 0 = [\tilde{F}_{H_0}]_{S^1}$ and $\Sigma_0 F_{H_j}$ is extendable to all previous isotropy subspaces and has an extension degree which is the product of 1 with the extension degree for F_{H_j}, that is 1. Thus, from Lemma 2.3, $\Sigma_0 F_{H_j}$ and $\tilde{F}_{H_j}$ are S^1-homotopic.

2) If $p \geq 1$, our hypotheses imply that $\Pi^{S^1}_{k+2(m+1)}(S^{l+2(n+1)})$ has just one component corresponding to $\mathbb{Z}_{m_0}$. Since one may have that $m_0 < l_0$, instead of looking at the isotropy subgroups, we shall compute the class of $\Sigma_r F$ with the detour through the K-degree.

Recall that, for $\Theta^*(F)$ (given by $z_j = Z_j^{m_j/l_0}$), one has:

$$deg_K(\Theta^*(F)) = \Pi_{j=1}^m (m_j/l_0) deg_E(F)$$

where Θ^*, deg_K and deg_E are morphisms. It is thus enough to compute the K-degree of $\Sigma_r F_d$ for the generator F_d given in Theorem 3.1, with $deg_E(F_d) = (-1)^k d$. Then

$$deg_K(\tilde{\Theta}^* \Sigma_r F_d) = \Pi_{j=1}^m (m_j/m_0)(q/m_0) deg_E(\Sigma_r F_d),$$

where

$$(\tilde{\Theta}^* \Sigma_r (F_d))(t, x_0, \tilde{Z}_1, \ldots, \tilde{Z}_m, Z) = (F_d(t, x_0, \tilde{z}_1^{m_1/m_0} \ldots, \tilde{z}_m^{m_m/m_0}), Z^{qr/m_0}).$$

Since $F_d(t, x_0, z_1, \ldots, z_{m-1}, z_m)$ has its zeros for $|z_m| = R/2$, then one may compute $deg_K(\tilde{\Theta}^*\Sigma_r F_d)$, by taking the section $0 \leq \tilde{Z}_m \leq R$, and $deg_K(\Theta^* F_d)$ by taking the section $0 \leq Z_m \leq R$. The counting argument of Theorem 3.1 will give that $\Theta^* F_d$ has $d\Pi_1^m m_j/l_0$ zeros, each with index $(-1)^k$. Thus

$$deg_K(\Theta^* F_d) = (-1)^k d\Pi_1^m (m_j/l_0).$$

For $\tilde{\Theta}^*\Sigma_r F_d$ one will get $d\Pi_1^{m-1} (m_j/m_0) m_m/l_0$ zeros, each with index $(-1)^k qr/m_0$ (coming from Z^{qr/m_0}). Thus

$$deg_K(\Theta^*\Sigma_r F_d) = (-1)^k d\Pi_1^m (m_j/m_0)(q/m_0)(m_0 r/l_0).$$

Then

$$deg_E(\Sigma_r F_d) = (m_0 r/l_0) deg_E(F_d)$$

and one has the result for general F.

Note that there are integers α, β such that $\alpha q + \beta l_0 = m_0$. Thus $m_0 r = \alpha qr + \beta r l_0$. Since qr is a multiple of m_m, it is also of l_0, thus $m_0 r$ is divisible by l_0. In particular, if $r = 1$ then $m_0 = l_0$.

Note also that if $p = 1$, then the map which has extension degree equal to $\Pi_1^n (n_j/m_j)(l_0/m_m)$ gives a map with extension degree $\Pi_1^n (n_j/m_j)(l_0/m_m)(m_0 r/l_0) = \Pi_1^n (n_j/m_j)(qr/q)(m_0/m_m)$, that is the group structure in Theorem 3.1 is preserved. Finally, for $p = 1$ and $n = 0$, one has $deg_E(F) = 0$.

3) For the case $k = l + 1$, if $[F]_{S^1} = \sum_0^s d_j [F_{H_j}]_{S^1}$ and $[\Sigma_r F]_{S^1} = \sum_0^{s+t} \tilde{d}_j [\tilde{F}_{H_j}]_{S^1}$, then $[\Sigma_r F]_{S^1} = \sum_0^s d_j [\Sigma_r F_{H_j}]_{S^1}$, since Σ_r and the extension degrees d_j are morphisms. It is thus enough to study $[\Sigma_r F_{H_j}]_{S^1}$.

a) If $j \in I_0$, that is $j = 0$ or $|H_j|$ doesn't divide qr, then the variables z and z^r play a dummy role. In this case $\Sigma_r F_{H_j} = F_{\tilde{H}_j}$.

b) If $j \in I_q$, that is $|H_j|$ divides q, then the complex dimensions of the isotropy subspaces are increased by one on each side. Let $V^{H_j} = \{(t, x_0, z_1, \ldots, z_u)\}$, with $m_1 \le m_2 \le \ldots \le m_u$ and $m_0 = |H_j| = (m_1 : \ldots : m_u)$. Then $q = \alpha m_0$ and $m_1 = \beta m_0$.

If $\beta = 1$ or $\alpha > 1$, then a look at the generator F_{H_j} is enough to see that $\Sigma_r F_{H_j}$ has an extension degree equal to r (use Remark 2.4, with the fact that the zeros of F_{H_j} are in $V^{H_j} \setminus \bigcup_{K > H} V^K$, property which is preserved for $\Sigma_r F_{H_j}$). Furthermore $\Sigma_r F_{H_j}$ and $r \tilde{F}_{H_j}$ have the same extension degree. Then, from Lemma 2.3 with $g(z_\perp) = z_j^{k_j}, j = u+1, \ldots, m$, both maps are S^1-homotopic on $S^{V \times \mathbb{C}}$, that is $[\Sigma_r F_{H_j}]_{S^1} = r[\tilde{F}_{H_j}]_{S^1}$.

If $\alpha = 1, q = m_0$ and $m_j > m_0$, we shall compute the extension degree of $\Sigma_r F_{H_j}$ by going first to Z_j with $z_j = Z_j^{m_j/m_0}, z = Z$. In fact, in order to compute directly the extension degree we should deform $\Sigma_r F_{H_j}$ to a map which would have to be non-zero for $z = 0$. However, from a look at F_{H_j}, one sees that $\Theta^*(\Sigma_r F_{H_j})$ has, for $0 \le Z_1 \le R$, $\Pi_{j=1}^u m_j/m_0$ zeros, each of index r. Thus the extension degree for $\Sigma_r F_{H_j}$ is r. Finally the same argument as above will give that $[\Sigma_r F_{H_j}]_{S^1} = r[\tilde{F}_{H_j}]_{S^1}$.

c) If $j \in I_{qr}$, that is $m_0 \equiv |H_j|$ divides qr but not q, then $(V \times \mathbb{C})^{H_j} = V^{H_j}$ but $(W \times \mathbb{C})^{H_j} = W^{H_j} \times \mathbb{C}$, then $\Sigma_r F_{H_j}$ has to extend to $(V \times \mathbb{C})^{H_j}$.

In fact, if $m_1 = m_0$, then F_{H_j} has its zeros for $|z_1| = R/2, z_s = 0, s \ge 2, t = 1/2$, $x_0 = 0$ $(x_0 = -R/2$ if $l = 0)$. Let

$$\tilde{F} = (F_{H_j}, z^r - K(t, x_0, z_1, \ldots, z_m)z_1^{qr/m_0})$$

with

$$K(t, x_0, z_1, \ldots, z_m) = 8t(1-t)(1 - |x_0|/R)\Pi_{s=1}^m (1 - |z_j|/R)(2/R)^{q/m_0}$$

(the factor $1 - |x_0|/R$ is replaced by $2(1 - |x_0|/R)$ if $l = 0$). Then $\tilde{F}|_{S^{H_j}} = \Sigma_r F_{H_j}|_{S^{H_j}}$ (since then $z = 0$ and $K = 0$), and is non-zero on B^{H_j}. The zeros of $\tilde{F}$ are those of F_{H_j} and $|z| = R/2$.

On the other hand if $m_s > m_0$ for $s = 1, \ldots, u$, then F_{H_j} has its zeros for $|z_s| = R/2$, $s = 1, \ldots, u, t = 1/2, x_0 = 0$ $(x_0 = -R/2$ if $l = 0)$. Now the map

$$\tilde{F} = (F_{H_j}, z^r - \tilde{K}(\Pi_1^u z^{\alpha_s})^{qr/m_0}),$$

with

$$\tilde{K} = K(2/R)^{q(u-1)/m_0} \text{ and } \sum_1^u \alpha_s m_s = m_0,$$

coincides with $\Sigma_r F_{H_j}$ on S^{H_j} and is non zero on B^{H_j}.

Both maps are non zero on $(V \times \mathbb{C})^K$ for any $K = \mathbb{Z}_{\alpha \bar{m}}$ with $\bar{m} = (q : m_0)$, and $\alpha > 1$, since the zeros have the isotropy group $\bar{H}_j = \mathbb{Z}_{\bar{m}}$.

Again in order to compute the extension degree on $(V \times \mathbb{C})^{\bar{H}_j}$, we shall go to

$$\Theta^* \tilde{F} = \tilde{F}(t, x_0, Z_1^{m_1/\bar{m}}, \ldots, Z_u^{m_u/\bar{m}}, \ldots, Z_v^{m_v/\bar{m}}, 0 \ldots, 0, Z^{q/\bar{m}}),$$

(one may have m_j a multiple of $\bar{m}$ but not of m_0), and compute the degree for $0 \le Z_1 \le R$. A simple count of zeros, in the four possible cases, will give a K-degree equal to

$$(\Pi_1^v m_s/\bar{m})(q/\bar{m})(\Pi_{u+1}^v k_s) r \bar{m}/m_0.$$

Thus the extension degree will be $q_j r \bar{m}/m_0$, that is $\Sigma_r F_{H_j}$ and $q_j r \bar{m}/m_0 \tilde{F}_{\bar{H}_j}$ have the same extension degree on $(V \times \mathbb{C})^{\bar{H}_j}$. Lemma 2.3 will imply that these two maps are S^1-homotopic on the sphere in $\mathbb{R} \times V \times \mathbb{C}$ and thus generate the same element.

Note again that one has two integers α and β such that $\alpha q + \beta m_0 = \bar{m}$, thus $r\bar{m} = \alpha qr + \beta m_0 r$, where qr is a multiple of m_0, that is $r\bar{m}/m_0$ is an integer.

Recall that one may have $[\tilde{F}_{\bar{H}_i}]_{S^1} = [\tilde{F}_{\bar{H}_j}]_{S^1}$: for example, take $m_1 = n_1 = 2$, $m_2 = n_2 = 3, q = 1, r = 6$, thus the isotropy subgroups for $\mathbb{R} \times V$ will be $H_0 = S^1$, $H_1 = \mathbb{Z}_3, H_2 = \mathbb{Z}_2, H_3 = \{e\}$ with $m_{H_1} = n_{H_1} = 1, m_{H_2} = n_{H_2} = 1, m_{H_3} = n_{H_3} = 2$. For the suspended problem one has the same isotropy subgroups but $I_0 = \{0\}, I_q = \{3\}$, $I_{qr} = \{1, 2\}, m_H < n_H$ for $H = H_1$ and $H_2, \bar{H}_1 = \bar{H}_2 = \bar{H}_3 = H_3$. Thus, taking $d_0 = 0$, $[\Sigma_r F]_{S^1} = 6d_3 [\tilde{F}_{H_3}]_{S^1} + 6((d_1/3) + d_2/2)[\tilde{F}_{H_3}]_{S^1} = (2d_1 + 3d_2 + 6d_3)[\tilde{F}_{H_3}]_{S^1}$. From this example it is clear that one may achieve a degree 1 (choosing $d_3 = 1, d_1 = d_2 = -1$) but then Σ_r cannot be one to one.

.

Note finally that if $q = |H_{j_0}|$, for some j_0, then q divides all modes corresponding to H_{j_0} and, if H_j is another isotropy subgroup, then $\bar{H}_j$ corresponds to the l.c.d. of the modes present for H_{j_0} and H_j, that is another isotropy subgroup for V. In this case there are no other isotropy subgroups for $V \times \mathbb{C}$ but those for V. If furthermore $r = 1$, then $I_{qr} = \emptyset$ and clearly Σ_1 is onto.

While if q doesn't divide any of the $m'_j s$, that is if $q \neq |H_j|$ for all j's, then $\mathbb{Z}_q$ is a new isotropy subgroup corresponding to the isotropy subspace $z_j = 0, j = 1, \ldots, m, z \neq 0$. On this subspace $\Sigma_r F_{H_j} \neq 0$ by construction and the corresponding extension degree would be 0. Thus, if $\tilde{F}_q$ is the generator for $\mathbb{Z}_q$, this generator or any multiple of it will not be achieved and Σ_r cannot be onto. Q.E.D.

Remark 3.2. In [**I.M.V**, Theorem B], we stated a general theorem for suspensions. In the case of Σ_0, the result ([**I.M.V**, Example B.1]) was that

Σ_0 *is an isomorphism if* $k + 2m_H \leq 2(l + 2n_H - 1)$ *for all isotropy subgroups* H *for* V.

For $k = l + 1 - 2p$, one has $2m_H \leq 4n_H + 2p + l - 3$ (thus $2p + l \geq 3$ for $H = S^1$). Thus, if $l \geq 3$, our condition $m_H \leq n_H + p$ is more restrictive. Furthermore if $p \geq 1$, then, since $k \geq 0$, $l \geq 1$ and the condition $2p + l \geq 3$ is met. Recall that in this case $m_H \leq n_H$ except if H is the smallest subgroup and then $m_H = m = n_H + p = n + p$, in which case the condition is $2n + l - 3 \geq 0$ which is violated only if $n = 0, l = 1$, thus $p = 1, m = 1$. But then the groups are 0 (see Theorem 3.1) and there is no gain. For $p = 0$ our results cover the case $l = 0$.

Note however that, from our explicit construction of the groups, we know where the isomorphism fails, for small l, and where the generators go for all l's.

On the other hand, for Σ_1, the conditions of [**I.M.V**, Theorem B] reduce to the following four assumptions:

1) $q = |H_0|$ for some isotropy subgroup of V (this is also a condition for Theorem 3.3 if $k = l + 1$ but not for $p \geq 1$),

2) if $|H|$ divides q, then $l + 2n_H \geq 2$, (since $k = l - 2p + 1 \geq 0$, this condition is violated only if $n_H = 0, p = 1 = l$ or $p = 0$ and $l < 2$),

3) for any isotropy subgroup H of V ($H = S^1$ is allowed), then $2(n_H - m_H + p) + 2n_H + l - 3 \geq 0$, (thus if $p = 0, l \geq 3$),

4) if K and H are isotropy subgroups, such that $|K|$ divides q and H is either S^1 or $|H|$ is a multiple of $|K|$ but q is not a multiple of $|H|$, then $n_K + p - m_H > 1$. (Note that these multiplicity assumptions imply that K is a strict subgroup of H).

Recall that the actions are such that $q, n_1, \ldots, n_{m-p}$ are multiples of m_r, for $r = m - p + 1, \ldots, m$ and that $n_j = k_j m_j$ for $j = 1, \ldots, m - p$ (thus q is a multiple of some $|H_0|$).

If $p \geq 1$ and $m = n+p$, then one has either $1 \leq m_H \leq n_H$ or $m_H = m = n_H + p = n + p$. Thus conditions (2) and (3) are met except, as already seen, if $n = 0, p = 1$ case which is excluded from our results. Condition (4), for $H = S^1$, reduces to $n_K + p > 1$ and could be violated only if $p = 1, n_K = 0 = n$, case excluded from Theorem 3.3. If H is not S^1, then $m_H < m_K$ and $n_K + p - m_H > n_K - m_K + p \geq 1$ unless $n_K = n, m_K = m$ and $|K| = (m_1 : \ldots : m_m)$. In this case the condition is $m - m_H > 1$. However if m_r is a multiple of $|H|$, for some $r = m - p + 1, \ldots, m$, then q will also be a multiple of $|H|$, that is one needs that $m_H \leq m - p$.

Thus the condition could be violated only if $p = 1$ and $|H| = (m_1 : \ldots : m_n)$. Now if one adds condition (1), which is stronger than ours, then $q = |H_0| = (m_{i_1} : \ldots : m_{i_j})$. If $m_{i_s} \neq m_m$ for all s, then q would be a multiple of $|H|$. If $m_{i_s} = m_m$, for some s, then, since q is a multiple of $m_m, q = m_m$, the other modes in H_0 (if any) would be multiples of m_m and one could have situations where $m_H = m - 1$ such as: $m_1 = 2, m_2 = 6, m_3 = 3 = q, n = 2$, then $l_0 = |K| = 1, |H| = (2 : 6) = 2$, where we know that Σ_1 is an isomorphism, but where the conditions of the general result don't apply. That is our present results, under the action hypotheses, are stronger, in particular for the case $p = 1$. (For $p > 1$, the only stronger condition would be (1)).

If $p = 0$, then $1 \leq m_H \leq n_H$ and since condition (3) implies $l \geq 3$ we have that conditions (1) and (2) are verified. The fourth condition implies that $n_{H_0} \geq 2$ (for $H = S^1$), which gives $n_K \geq 2$ for any K with q a multiple of $|K|$.

On the other hand one may construct many examples where $n_K - m_H = 1$, (for instance $|H_0| = q = (2 : 6) = 2, |K| = (2 : 6 : 9) = 1, |H| = (6 : 9) = 3$). Thus this condition is much stronger than ours and could be met if for each mode one has at least two components with the same mode.

Remark 3.3. In order to define the S^1–degree for infinite dimensional spaces, one needs only that Σ_1 is one to one, since then $\Sigma_1 F$ will have the same components $d_0, \ldots, d_s$ and zero on the others (if any). This fact will be used in chapter four.

In this case the arguments of [**I.M.V**, Proposition 3.1] apply. Furthermore for the additivity of the degree, we need that Σ_0 is a monomorphism [**I.M.V**, Corollary 2.5], thus one will have problems only if $l = 2, k = 3$ and only with the invariant part of the maps. Finally, one of our reasons for introducing the degree as an obstruction to extension is that if Ω is a ball and the degree of f is zero, then f has a non-zero extension to Ω. Since $\deg_{S^1}(f; \Omega) = \Sigma_0[f]_{S^1}$, [**I.M.V**, Remark 2.3], then, if f has zero degree, f will be trivial except if $l = 3, k = 4$, where one needs to have the triviality of $f|_{V^{S^1} \cap \partial\Omega}$ which is given by an element in $\Pi_{k-1}(S^{l-1})$. In all other cases Σ_0 is one to one. Again, this information will be used in the next chapter where the S^1-degree will be studied.

3.4. Relationship with the set of K-degrees.

For each isotropy subgroup H, one may compute $[\tilde{F}^H]_{S^1} = \Theta_H^*([F]_{S^1}) = [F^H(t, x_0, Z_j^{m_j/m_0})]$, where one considers only Z_j's such that m_j is a multiple of $m_0 = |H|$. This class is the K-degree of F^H and was studied in [**I.M.V**].

If $p \geq 1$, our hypotheses imply that $\Theta_H^*([F]_{S^1}) = 0$, unless H is the smallest isotropy subgroup and then

$$deg_K(\tilde{F}^H) = \Pi(m_j/m_0)deg_E(F^H).$$

Consider then the case $k = l + 1$. Since $[F]_{S^1} = \sum_0^s d_j[F_{H_j}]_{S^1}$, it is enough to compute $\Theta_H^*(F_{H_j})$ for each $j \geq 1$ (the invariant part of F gives the first part of $\Theta_H^*[F]_{S^1}$).

Now if H is not a subgroup of H_j, then $|H|$ doesn't divide $|H_j|$. Let $|H_j| = (m_1 : \ldots : m_r)$. If $m_1 = |H_j|$, then $V^H \subset \{z_1 = 0\}$ and $F_{H_j}|_{V^H} \neq 0$ in the two cases of generators defined in the proof of Theorem 3.2. If none of the $m_1, \ldots, m_r$ is equal to $|H_j|$, then one may take m_1 such that m_1 is not a multiple of $|H|$ which implies that $V^H \subset \{z_1 = 0\}$ and thus $F_{H_j}|_{V^H} \neq 0$. Hence $\Theta_H^*(F_{H_j}) = 0$ in this case.

If $H < H_{j_0}$, let $m_0 = |H|, m_s = |H_{j_0}|$, thus m_0 divides m_s. As in the proof of Theorem 3.2, one has to compute the K–degree of $F_{H_{j_0}}^H$. A look of the four possible generators will be enough to see that this K–degree (for $0 \leq Z_1 \leq R$) is $(m_1/m_s)\Pi_1^r(m_j/m_0)\Pi(n_j/m_0)$, where the last product in over all j's such that m_j is a multiple of m_0 but not of m_s.

Thus we have proved the following result:

Theorem 3.4.

1) If $k \leq l - 1$, then $\Theta_H^*[F]_{S^1} = 0$ if H is not the smallest isotropy subgroup. While, if H is the smallest subgroup, $\Theta_H^*[dF_H] = d\Pi(m_j/|H|)$, when F_H has extension degree 1.

2) If $k = l + 1$ and:

a) if $m_H < n_H$, then $\Theta_H^*[F]_{S^1} = 0$;

b) if $m_H = n_H$, then :

$$\Theta_H^*(\sum_0^s d_j[F_{H_j}]_{S^1}) = (d_0[F_{H_0}], \sum_{H \leq H_{j_0}} d_{j_0}|H|/|H_{j_0}|\Pi_{J_{H_{j_0}}}(m_j/|H|)\Pi_{J_H \backslash J_{H_{j_0}}}(n_j/|H|).$$

where $J_H = \{j, \text{ such that } m_j \text{ is a multiple of } |H|\}$.

Remark 3.4. From Theorem 3.4, one sees immediately that if $\Theta_H^*[F]_{S^1} = 0$ for all H's, then $[F]_{S^1} = 0$. In fact the assignment is representable as a triangular matrix with diagonal terms $\alpha_H = \Pi_{J_H}(m_j/|H|)$. Thus two maps will have the same S^1- degree if and only if they have the same almost-semi- free degrees, as defined in **[I.M.V]**, for all H's. Note however that one may easily construct examples to show that this assignment is not onto, i.e., that this matrix is not invertible over the integers: in fact one would need that $\Pi\alpha_{H_j} = 1$, thus $m_j = |H|$ for all j in J_H and all H's. This is possible only if there is just one mode. The triangular matrix will be studied further in Chapter V.

Remark 3.5. If one takes one of the components of $\mathbb{C}^n$ to some power β, as done in **[I.M.V**, Section 4.1], then d_j is replaced by βd_j if the component belongs to W^{H_j} (look at the generators). Furthermore if one forgets the group action, i.e., if one takes χ_* in the sense of **[I.M.V**, Corollary 4.7], then for $k = l + 1$, one gets for each $F_{H_j}|V^{H_j}$ an element in $\mathbb{Z}_2$ of degree equal to the parity of $\sum_{s=1}^r n_s/|H_j| - \sum_{s=2}^r m_s|H_j|$, (see the generators of Theorem

3.2 and the proof of Corollary 4.7 in [I.M.V]; the case $p \geq 1$ was already treated in our previous paper).

Remark 3.6. If H is an isotropy subgroup, the restriction of a map F to V^H, i.e., the application $F \to F^H$, clearly induces a morphism:

$$\Pi_{k+2m}^{S^1}(S^{l+2n}) \to \Pi_{k+2m_H}^{S^1}(S^{l+2n_H}).$$

If $k = l + 1$, a look at the generators given in Theorem 3.2 is enough to see that if

$$[F]_{S^1} = \sum d_j [F_{H_j}]_{S^1},$$

then

$$[F^H]_{S^1} = \sum_{H_j \geq H} d_j [F_{H_j}^H]_{S^1}.$$

Thus the restriction is an epimorphism. In particular if $F^H \neq 0$ then $d_j = 0$ for all $j's$ such that $H_j \geq H$.

3.5. Symmetry Breaking.

In some problems, one may force or perturb an equivariant function by another function which may not have the full equivariance. In the present case, a S^1-problem perturbed by any function will lose its symmetry. The unperturbed problem will have a S^1-degree, while the forced problem will have an ordinary degree (that is without action) which, for small forcing, will be the ordinary degree of the unperturbed problem. It is thus of interest to study the image of $\Pi_{k+2m}^{S^1}(S^{l+2n})$ in $\Pi_{k+2m}(S^{l+2n})$, under the natural homomorphism:

$$\chi_* : \Pi_{k+2m}^{S^1}(S^{l+2n}) \to \Pi_{k+2m}(S^{l+2n})$$

$$\chi_*([F]_{S^1}) = [F].$$

χ_* is clearly a morphism since the addition is defined on the t variable.

If $k = l, m = n$ and $m_j = k_j n_j$, we have seen in Corollary 2.2 and Remark 2.3, that:

$$\chi_*([F]_{S^1}) = \chi_*([(F^{S^1}, z_1^{k_1}, \ldots, z_n^{k_n})]_{S^1})$$
$$= \Pi k_j [F^{S^1}] \in \mathbb{Z}.$$

If $k = l + 1 - 2p$ and $n = m - p$, then $\Pi_{k+2m}(S^{l+2n}) = \mathbb{Z}_2$, unless $n = 0, m = 1, l = 1$ or 2, or $n = m = 1, l = 0$, cases which are not too interesting.

If $p \geq 1, n_j = k_j m_j$ for $j = 1, \ldots, n, n_j$ is a multiple of $m_{n+1}, \ldots, m_m$ and $m_m = (m_1 : \ldots : m_m)$, we have seen in Remark 3.1 that:

$$\text{if } [F]_{S^1} = d \in \mathbb{Z}, \text{ then } \chi_*([F]_{S^1}) \equiv d(\sum_1^n (k_j - 1)\frac{m_j}{m_0} + \sum_{n+1}^m \frac{m_j}{m_0} + 1).$$

If $k = l + 1, m = n$ and $n_j = k_j m_j$, then for each isotropy subgroup H_s, let I_s be the set of m_j's which are multiples of $|H_s|$ corresponding to the modes present in V^{H_s}. One has then the following result:

Theorem 3.5. *If* $[F]_{S^1} = \sum d_s[F_{H_s}]_{S^1}$, *then*

$$\chi_*([F]_{S^1}) \equiv \sum_s d_s(\sum_{j \in I_s}(k_j - 1)m_j/|H_s| + 1)(\Pi_{j \notin I_s} k_j),$$

in particular if $k_j \equiv 1$, *then* $\chi_*([F]_{S^1}) \equiv \sum_s d_s$.

Proof. Since χ_* is a morphism, it is enough to compute $\chi_*([F_{H_s}]_{S^1})$, for each of the generators defined in the proof of Theorem 3.2.

For $H_s = S^1$, then $I_s = \emptyset$ and, from [**W**, p.479], $\chi_*([F_0]_{S^1}) = \Pi k_j$.

If $|H_s| = m_1$, then, if $l \geq 1$:

$$F_{H_s} = (1 - 2|z_1|/R, 2t - 1, \tilde{x}_0, \lambda z_1^{k_1}, \{z_1^{n_j/m_1} \bar{z}_1^{m_j/m_1} z_j\}_{j \in I_s \setminus \{1\}}, \{z_j^{k_j}\}_{j \notin I_s})$$

$x_0 = (x, y, \tilde{x}_0), \lambda = x + iy$, and if $l = 0$:

$$F_{H_s} = (1 + 2x_0/R + 2(2|z_1| - R)^2/R^2, (2t - 1 - i(2|z_1| - R))z_1^{k_1},$$
$$\{z_1^{n_j/m_1} \bar{z}_1^{m_j/m_1} z_j\}_{j \in Is \setminus \{1\}}, \{z_j^{k_j}\}_{j \notin I_s}).$$

If $|H_s| > m_j$ for all j's in I_s, then the expressions for F_{H_s} are given in the proof of Theorem 3.2 (γ) and (δ). For the moment the explicit form will not be necessary: in fact, let us suspend F_{H_s} by z with action $e^{iq\varphi}$ in the domain and range, where $q = |H_s|$. From Theorem 3.3 (3), this suspension is an isomorphism and $[\Sigma_1 F_{H_s}]_{S^1} = [\tilde{F}_{H_s}]_{S^1}$. On the other hand, $\chi_*([\Sigma_1 F_{H_s}]_{S^1}) = \Sigma^2 \chi_*([F_{H_s}]_{S^1})$ and, since $n \geq 2$ (I_s has at least two elements),

Σ^2 is an isomorphism. Thus, one may replace F_{H_s} by $\tilde{F}_{H_s}$ and assume that $|H_s| = m_1, I_s$ is increased by one element with a corresponding $k_1 = 1$. Thus, the formula $\sum (k_j - 1) m_j / |H_s|$ will be unchanged.

If $|H_s| = m_1$ and $l \geq 1$, the term Πk_j will come from [**W**, p.479] and, after suspensions, one has to consider the class of

$$\left(1 - 2|z_1|/R, \lambda z_1^{k_1}, \{ z_1^{n_j/m_1} \bar{z}_1^{m_j/m_1} z_j \}_{j \in Is \setminus \{1\}} \right).$$

The non-equivariant rotation:

$$\begin{pmatrix} \tau & 1 - \tau \\ -(1 - \tau) & \tau \end{pmatrix} \begin{pmatrix} \tau z_1^{k_1} & -(1 - \tau) \\ (1 - \tau) z_1^{k_1 + n_j/m_1} \bar{z}_1^{m_j/m_1} & \tau z_1^{n_j/m_1} \bar{z}_1^{m_j/m_1} \end{pmatrix} \begin{pmatrix} \lambda \\ z_j \end{pmatrix}$$

is an admissible deformation since, on a zero of the map, one has $|z_1| = R/2$. Using again the suspension theorem, one is reduced to the class of $(1 - 2|z_1|/R, \lambda z_1^\alpha)$, with $\alpha = k_1 + \sum_{I_s \setminus \{1\}} (n_j - m_j)/m_1 = \sum_{I_s} (k_j - 1) m_j / m_1 + 1$. The last map is $\alpha \eta$, where the Hopf map η generates $\Pi_3(S^2)$.

If $|H_s| = m_1$ and $l = 0$, the same rotations will reduce the study to

$$(1 + 2x_0/R + 2(2|z_1| - R)^2/R^2, (2t - 1 - i(2|z_1| - R)) z_1^\alpha)$$
$$= (a + 2b^2, (2t - 1 - iRb) z_1^\alpha).$$

The deformation $(\tau(a + 2b^2) - (1 - \tau) bR, (2t - 1 - i(\tau bR + (1 - \tau)(a + 2b^2)) z_1^\alpha)$ is easily seen to be valid (if $z_1 = 0$ then $b = -1$ and $a \geq -1$ on the ball, for $z_1 \neq 0$ one has performed a rotation on the terms $a + 2b^2$ and bR).

For $\tau = 0$, the map $(R - 2|z_1|, (2t - 1 - i(1 + 2x_0/R + 2(2|z_1| - R)^2/R^2)) z_1^\alpha)$ is clearly deformable to $(R - 2|z_1|, (2t - 1 - i(1 + 2x_0/R)) z_1^\alpha)$ and to $(R - 2|z_1|, (2t - 1 - i x_0) z_1^\alpha)$ which has the class of $\alpha \eta$.

Q.E.D.

Remark 3.7. If $|H_s| > m_j$, one may also compute the class $\chi_* ([F_{H_s}]_{S^1})$ directly with the following argument. For $l \geq 1$, the generator was defined as:

$$F_{H_s} = (1 - 2|z_1|/R, 2|z_2|\dots|z_r|t - (R/2)^{r-1}, \tilde{x}_0, \lambda z_1^{k_1},$$
$$\{((\Pi_{q=1}^{j-1}\bar{z}_q^{\alpha_{q,j}-1})^{m_j/l_j} z_j^{l_j-1/l_j} - \epsilon_j)z_j^{k_j}\}_{j\in I_s\setminus\{1\}}, \{z_j^{k_j}\}_{j\notin I_s}),$$

where $\sum_q \alpha_{q,j} m_q = l_j$.

Through a series of rotations one may replace $z_1^{k_1}$ by $(\Pi z_j^{k_j})_{j\in I_s}$, deform $2|z_2|\dots|z_n|t - (R/2)^{r-1}$ to $2t - 1$ and, again via rotations and scaling, consider the map defined for $|z_j| \le 2, |\lambda| \le 2$:

$$\left(1 - |z_1|, \lambda z_1^{\alpha_1}\dots z_n^{\alpha_n}, \{z_j^{l_j-1/l_j} - (\Pi\bar{z}_q^{\alpha_{q,j}-1})^{m_j/l_j}\}_{j\in I_s\setminus\{1\}}\right),$$

with $\alpha_j = k_j - \sum \alpha_{j,q} m_{q+1}/l_{q+1}$, and $\alpha_{j,q} = 0$ if $j > q$ and $m_{r+1} = 0$.

This map has its zeros at $\lambda = 0, |z_j| = 1$. Write the map as:

$$(1 - |z_1|, \lambda\eta_1, \eta_2, \dots, \eta_r).$$

Note that if $\eta_r = \dots = \eta_2 = 0$, then it is easy to show, by induction and from the definition of l_{j-1}, that $|z_j| = |z_1|^{m_j/m_1}$. In particular the only zero of the transformation $(z_1, \dots, z_r) \to (\eta_1, \dots, \eta_r)$ is at the origin. Furthermore the deformation $\tau(1 - |z_1|) + (1 - \tau)(1 - |\eta_1|)$ is valid (one may write this expression as $(1 - |z_1|)(\tau + (1 - \tau)\sum_1^\beta |z_1|^{\beta-j})$, on a zero of $\eta_2, \dots, \eta_r$ with $\beta = \sum |\alpha_j| m_j/m_1$).

Thus the class of $\chi_*([F_{H_s}]_{S^1})$ is $(\Pi_{j\notin I_s} k_j) M \sum^{l+2(n-1)} \eta$, where M is the degree of the transformation: $(z_1, \dots, z_r) \to (\eta_1, \dots, \eta_r)$.

Going through the detour $z_j = Z_j^{m_j}$, M will be the index, at 0, of the transformation:

$$(Z_1, \dots, Z_r) \to (Z_1^{m_1\alpha_1}\dots Z_r^{m_r\alpha_r}, \{Z_j^{m_j l_j-1/l_j} - (\Pi_1^{j-1} Z_q^{\alpha_{q,j}-1 m_q})^{m_j/l_j}\})$$

divided by Πm_j. The zeros of the last $r - 1$ components are for $|Z_j| = |Z_1|$ for all j's.

In order to compute the index, one may perturb $Z_r^{m_r\alpha_r}$ to $Z_r^{m_r\alpha_r} - 1$, (we are working in the set $|Z_j| < 2$), generating $m_r\alpha_r$ additional strings of zeros with $|Z_j| = 1$. One has then to compute the index at the origin of the new map, together with the degree in a neigborhood of the torus $|Z_j| = 1$.

Near the origin, one may deform the first component to $-Z_1^{m_1\alpha_1}\dots Z_{r-1}^{m_{r-1}\alpha_{r-1}}$ and the last component to $Z_r^{m_r l_{r-1}/l_r}$. The contribution of this part will be $(m_r l_{r-1}/l_r)I_{r-1}$, where I_{r-1} is the index at the origin of the original map without Z_r and without the last component.

Near the torus, one has to count the number of zeros and to compute their index. In order to do so one may use the transformation:

$$(Z_1,\dots,Z_r) \to (Z_1, Z_2\bar{Z}_1,\dots,Z_r\bar{Z}_1) = (Z_1,\xi_2,\dots,\xi_r),$$

which is an orientation preserving homeomorphism near the torus.

the transformed map is

$$(Z_1^\alpha, \xi_2^{m_2\alpha_2}\dots\xi_{r-1}^{m_{r-1}\alpha_{r-1}}(\xi_r^{m_r\alpha_r} - \bar{Z}_1^{m_r\alpha_r})/|Z_1|^{2(\alpha-m_1\alpha_1)}, Z_1^{m_j l_{j-1}/l_j} A_j/|Z_1|^{2m_j l_{j-1}/l_j})$$

where $\alpha = \sum d_j m_j$ and $A_j = (\xi_j^{m_j l_{j-1}/l_j} - (\Pi_2^{j-1}\xi_1^{\alpha_{q,j-q}m_q})^{m_j/l_j}|Z_1|^{2\alpha_{1,j-1}m_1 m_j/l_j})$.

The zeros of the last $r-1$ components are for $|\xi_j| = |Z_1|^2$ and, for $|Z_1| \neq 0$, one has $(\Pi_1^r m_j)/l_r$ different zeros. In the torus, $|\bar{Z}_1| = |\xi_r| = |Z_1|^2$, thus $|Z_1| = 1$ and one obtains $m_r|\alpha_r|$ possible phases for Z_1 for any given ξ_r: the number of zeros on the torus is $(\Pi_1^r m_j)m_r|\alpha_r|/l_r$.

Near one of these zeros $(Z_1^0, \xi_2^0,\dots,\xi_r^0)$, one may deform the terms $Z_1^\alpha, \xi_j^{m_j\alpha_j}, Z_1^{m_j l_{j-1}/l_j}$ to the corresponding value at the zero and then to 1. One may also replace recursively $(\Pi_2^{j-1}\xi_q^{\alpha_{q,j-1}m_q})^{m_j/l_j}|Z_1|^{2\alpha_{1,j-1}m_1 m_j/l_j}$ by $(\Pi_2^{j-1}\xi_q^{\alpha_{q,j-1}m_q})^{m_j/l_j}|Z_1|^{2m_j/l_j} = \xi_j^{0m_j/l_j}|Z_1|^{2m_j/l_j}$.

One is then reduced to compute the index at the above zero of

$$(\xi_r^{m_r\alpha_r} - \bar{Z}_1^{m_r\alpha_r}, \{\xi_j - \xi_1^0|Z_1|^2\}_{j=2,\dots r-1}).$$

One may deform the first component to $\bar{Z}_r^{m_r\alpha_r}(\xi_r^{0m_r\alpha_r}Z_1^{m_r\alpha_r} - 1)$ then to $Z_1^{m_r\alpha_r} - Z_1^{0m_r\alpha_r}$ and finally to $Z_1 - Z_1^0$, if $\alpha_r > 0$, obtaining an index equal to 1, to $\bar{Z}_1 - \bar{Z}_1^0$ if $\alpha_r < 0$ with an index -1, or to $|Z_1|^2 - 1$ if $\alpha_r = 0$ with index 0.

Hence $I_r = I_{r-1}m_r l_{r-1}/l_r + m_r\alpha_r(\Pi_1^r m_j)/l_r$.

$$I_r = [(\Pi_{j+1}^r m_q) l_j I_j + (\sum_{j+1}^r \alpha_q m_q) \Pi_1^r m_q]/l_r.$$

Since I_1 is the index at 0 of $Z_1^{m_1 \alpha_1}$, one has

$$I_r = (\sum_1^r \alpha_j m_j)(\Pi_1^r m_j)/l_r.$$

The index of the transformation $(z_1, \ldots, z_r) \to (\eta_1, \ldots, \eta_r)$ is then

$$(\sum_1^r \alpha_j m_j)/l_r = \sum (k_j m_j - \sum \alpha_{j,q} m_j m_{q+1}/l_{q+1})/l_r$$
$$= (\sum k_j m_j - \sum m_{q+1} l_q/l_{q+1})/l_r.$$

Thus the index of the transformation is, with $l_0 \equiv 0$:

$$\sum (k_j - l_{j-1}/l_j) m_j/l_r.$$

It remains to show that $\sum (\tilde{l}_{j-1}/\tilde{l}_j) \tilde{m}_j$ and $\sum \tilde{m}_j$, where $\tilde{l}_j = l_j/l_s, \tilde{m}_j = m_j/l_s$, have opposite parities. Since $\tilde{m}_j = (\tilde{l}_{j-1} : \tilde{m}_j)$, if $\tilde{l}_j$ is even, so are $\tilde{l}_{j-1}$ and $\tilde{m}_j$; if $\tilde{l}_{j-1}$ and $\tilde{l}_j$ are odd, then $(\tilde{l}_{j-1}/\tilde{l}_j)\tilde{m}_j \equiv \tilde{m}_j$; if j_0 is the first index for which l_{j_0-1} is even and l_{j_0} is odd, then $\tilde{m}_{j_0}$ is odd and $(\tilde{l}_{j_0-1}/\tilde{l}_{j_0}) \tilde{m}_{j_0} \equiv \tilde{m}_{j_0} + 1$ and one has proved this final claim.

The case $l = 0$ will be left to the reader.

This symmetry breaking argument will be applied in the next chapters to ordinary differential equations.

Remark 3.8. If $\chi_*([F]_{S^1}) = 0$, then one may have a non equivariant perturbation of F without zeros. For example, the map

$$(1 - 2(|z_1| + |z_2|))/R, 2t - 1, \lambda z_1, \lambda z_2), \text{ with } \lambda = x + iy, k_1 = k_2 = 1,$$

has S^1-degree (0,2), since it is deformable to $(1 - 2(|z_1| + |z_2|))/R, 2t - 1, \lambda^2 z_1, z_2)$, via the equivariant rotation $\begin{pmatrix} (1-\tau) & \tau \\ -\tau & (1-\tau) \end{pmatrix} \begin{pmatrix} (1-\tau)\lambda & \tau \\ \tau\lambda^2 & (1-\tau)\lambda \end{pmatrix} \begin{pmatrix} z_1 \\ z_2 \end{pmatrix}.$

However the non-equivariant perturbation $(\lambda z_1 - t\bar{z}_2, \lambda z_2 + t\bar{z}_1)$ has no zeros for any $t \neq 0$.

CHAPTER FOUR

DEGREE OF S^1 - MAPS

Let Ω be an open bounded invariant subset of $\mathbb{R}^k \times \mathbb{C}^m$ and $f(x_0, z)$ be an equivariant map with values in $\mathbb{R}^l \times \mathbb{C}^n$ and non zero on $\partial\Omega$. Then, if $\tilde{f}(x_0, z)$ is an equivariant extension of f to an invariant ball B containing Ω,

$$\deg_{S^1}(f; \Omega) = [2t + 2\phi(x_0, z) - 1, \tilde{f}(x_0, z)]_{S^1}$$

where $\phi = 0$ in $\bar{\Omega}$ and $\phi = 1$ outside an invariant neighborhood of Ω.

If $k = l, m = n$ and $m_j = k_j n_j$, then, from Corollary 2.2 and Remark 2.3, one has that:

$$\begin{aligned} \deg_{S^1}(f; \Omega) &= [2t + 2\phi(x_0, 0) - 1, \tilde{f}(x_0, 0)] \\ &= \deg(f^{S^1}; \Omega^{S^1}) \end{aligned}$$

as remarked in the introduction. In this case one also has:

$$\deg(f; \Omega) = (\Pi k_j) \deg(f^{S^1}; \Omega^{S^1}).$$

Thus, the information about f is completely contained in the invariant part. This is a Borsuk-Ulam result for the S^1-action.

We shall then consider the situation where one has results which depend on the equivariance, and in particular for the case when $k = l + 1 - 2p$.

As we have already seen $\deg_{S^1}(f, \Omega)$, as an element of $\Pi^{S^1}_{k+2m}(S^{l+2n})$, has all the properties of the usual degrees, with the precautions indicated in Remark 3.3 for the additivity if $l = 2, k = 3$ and for the extension property if $l = 3, k = 4$. Our hypotheses on the actions are still those of Chapter 3.

Our goal in this chapter is first to characterize completely the S^1-degree by giving its range in $\Pi_{k+2m}^{S^1}(S^{l+2n})$, then to extend it to the infinite dimensional setting and finally to compare it to other degrees.

4.1. Range of $\deg_{S^1}(f; \Omega)$.

Given the above hypotheses one has the following result.

Theorem 4.1.

1) *If $k \leq l - 1$, then any element of $\Pi_{k+2m}^{S^1}(S^{l+2n})$ is achieved as the S^1-degree of a map defined on $\bar{\Omega}$.*

2) *If $k = l + 1$, then any element $\sum d_j[F_{H_j}]_{S^1}$ in $\Pi_{k+2m}^{S^1}(S^{l+2n})$ is achieved by the S^1-degree of a map f, provided $d_j = 0$ if $\Omega^{H_j} = \emptyset$ and $d_0 = 0$ if $l = 2$. If $l = 0$, then a necessary condition for the existence of a non-zero S^1-map on $\partial\Omega$ is that $\bar{\Omega}^{S^1} = \emptyset$.*

3) *In all cases the degree is fully additive (restricting to maps with $d_0 = 0$ if $l = 2$), and for the Hopf property ($\Omega = B$) one needs to check that f^{S^1} is also trivial in the case $l = 3, k = 4$.*

Proof: 1) Since Ω is open then there is a point (x^0, z^0) in Ω such that $|z_j^0| \equiv |\eta_j| > 0$. One may assume, by translation of coordinates, that $x^0 = 0$. Consider the map:

$$f(x_0, z) = (|\eta_m| - |z_m|, x_0, ((\Pi_1^{j-1} \bar{z}_q^{\alpha_{q,j}-1} \bar{z}_m^{\alpha_{m,j}-1})^{m_j/l_j} z_j^{l_j-1/l_j} - \varepsilon_j)^{\gamma_j} z_m^{n_j \delta_j/m_m})$$

where j runs from 1 to $m - 1$, $\alpha_{q,j}$ are defined as in Theorem 3.1. If $p > 1$, then $\gamma_j = (-1)^l d$ if $j = n + 1$ and $\gamma_j = 1$ for $j \neq n + 1$, while if $p = 1$, then $\gamma_1 = (-1)^l d$ and $\gamma_j = 1$ for $j \geq 2$. One takes $\delta_j = 1$ if $1 \leq j \leq n$ and $\delta_j = 0$ if $j \geq n + 1$. Finally ε_j are chosen such that the only zeros of f are for $|z_j| = |\eta_j|$.

Note that if $0 \leq z_m \leq R$, then $f(x_0, z)$ has m_m/m_0 zeros, each with index d. One may choose the phases of ε_j in such a way that they give the phases of z_j^0 for one of these zeros. Now the orbit of $(z_1^0, \ldots, z_m^0)$ has isotropy type m_0 and cuts m_m/m_0 times the semi-hyperplane $0 \leq z_m \leq R$. Hence all the zeros of $f(x_0, z)$ are on the same orbit and thus they belong to Ω. One may then choose $\tilde{f}(x_0, z) = f(x_0, z)$, the term $\phi(x_0, z)$ is deformable to

0 and, as in the proof of Theorem 3.1, f has degree d in $\Pi_{k+2m}^{S^1}(S^{l+2n})$, (the change from $(-1)^k$ to $(-1)^l = -(-1)^k$ comes from the fact that the t component in Theorem 3.1 was in the second place and not the first).

2) If $k = l + 1$ and $\bar{\Omega}^H = \emptyset$, then one may choose $\phi(x_0, z)$ such that $\phi = 1$ in a neighborhood of V^H. From the explicit construction of the generators, it is clear that if $[F]_{S^1} = \sum d_j[F_{H_j}]_{S^1}$, then $[F^H]_{S^1} = \sum_{H \leq H_j} d_j[F_{H_j}^H]_{S^1}$. Then $2t + 2\phi - 1 = 2t + 1$ on V^H, $[F^H]_{S^1} = 0$ and the corresponding d_j's are 0 (recall that if $H \leq H_j$ then $V^{H_j} \subset V^H$).

Consider now the case where $\Omega^H \neq \emptyset$ and, of course, $m_H = n_H$. Let (x_0^0, z^0) be in Ω^H, which is open in V^H. One may thus assume, as before, that $x_0^0 = 0$ and $|z_j^0| \equiv |\eta_j| > 0$.

a) If $l \geq 1$ and there is one mode equal to $|H|$, let us say $m_1 = |H|$, then consider the map:

$$f_{(-1)^l d, H}(x_0, z) = (|\eta_1| - |z_1|, \tilde{x}_0, \lambda^d z_1^{k_1}, z_1^{n_j/m_1}(\bar{z}_1^{m_j/m_1} z_j - \varepsilon_j)_{j=2,\ldots,r}, z_{r+1}^{k_{r+1}}, \ldots)$$

where $x_0 = (x, y, \tilde{x}_0)$, $\lambda = x + iy$, $|\varepsilon_j| = |\eta_j|/|\eta_1|^{m_j/m_1}$, $k + 2r = \dim V^H$.

Again $\phi(x_0, z)$ may be deformed to 0 and one obtains that $\deg_{S^1}(f_{(-1)^l d, H}; \Omega) = (-1)^k d[F_H]_{S^1}$, as given in Theorem 3.2. (Note again the change of orientation in the range with respect to the t component and that there is just one orbit of zeros).

b) If $l \geq 1$ and $m_1 = s|H|$, consider the map:

$$f_{(-1)^k d, H}(x_0, z) = \left(|\eta_1|^\beta \Pi_{j=1}^r |\eta_j| - |z_1|^\beta \Pi_{j=1}^r |z_j|, \tilde{x}_0, \lambda^d z_1^{k_1}, \right.$$
$$\left. z_j^{k_j}\left((\Pi_1^{j-1} \bar{z}_q^{\alpha_{q,j-1}})^{m_j/l_j} z_j^{l_{j-1}/l_j} - \varepsilon_j\right)_{j=2,\ldots,r}, z_{j+r}^{k_{j+r}}, 0, \ldots, 0\right)$$

where $\beta > 0$, ε_j are given in such a way that the zeros of this map are for $|z_j| = |\eta_j|$. In fact the zeros of this map are for $z_j \neq 0$ $j = 1, \ldots, r$, thus $\Pi_1^{j-1} |z_q|^{|\alpha_{q,j-1}| m_j/l_j} |z_j|^{l_{j-1}/l_j} = |\varepsilon_j|$ (recall that z^α means $\bar{z}^{|\alpha|}$ if $\alpha < 0$). Choose $|\varepsilon_j|$ such that the solutions of this recursive set are of the form $|z_j| = |\eta_j||z_1/\eta_1|^{\beta_j}$, where β_j is some rational number ($\beta_2 = -m_2/l_1$, etc...). Then the first component will give: $|\eta_1|^{\beta+1} - |z_1|^{\beta+1}||z_1/\eta_1|^{\sum_2^r \beta_j}$. Choose $\beta \geq 0$ such that $\beta + \sum_{2}^{r} \beta_j \geq 0$. Then the unique zero will be for $|z_1| = |\eta_1|$. The phases of ε_j are chosen such that the s zeros of f, for $0 \leq z_1 \leq R$, each of index d, correspond to the orbit of $(z_1^0, \ldots, z_m^0)$, thus, these zeros are in Ω.

Now the maps $2t + 2\phi - 1$ and $2t\Pi_1^r|z_j| - \Pi_1^r|\eta_j|$ are linearly deformable one to the other on the set of zeros of $f_{(-1)^k d, H}$.

Furthermore one may perform linear deformations in $|\eta_j|$ or in $|z_j|$ in the first component of the above map (together with the condition $2t\Pi|z_j| - \Pi|\eta_j|$): replace $|\eta_j|$ by $|z_j||z_1/\eta_1|^{-\beta_j}$ if $\beta_j \leq 0$ and replace $|z_j|$ by $|\eta_j||z_1/\eta_1|^{\beta_j}$ if $\beta_j > 0$.

One will get:

$$\Pi_{\beta_j \leq 0} \left(|z_j||z_1/\eta_1|^{-\beta_j}\right) \left(\Pi_{\beta_j > 0}|\eta_j|\right) |\eta_1|^{\beta+1} \left(1 - |z_1/\eta_1|^{\beta+1+\sum \beta_j}\right).$$

It is clear that this expression (together with the condition) can be deformed to:

$$\left(2t\Pi_1^r|z_j| - \Pi_1^r|\eta_j| \quad , \quad |\eta_1| - |z_1|\right).$$

This pair is deformable to $(2t\Pi_2^r|z_j|\Pi_2^r|\eta_j||\eta_1| - |z_1|vert)$ and one has again a S^1-degree which is $(-1)^l d[F_H]_{S^1}$.

c) If $l = 0$, then from the equivariance one has $f(x_0, 0) = 0$ and one could not have a non-zero map on $\partial \Omega^{S^1}$, unless $\bar{\Omega}^{S^1} = \emptyset$. Let then $(0, z_1^0, \ldots, z_r^0)$ be in Ω^H, with $|z_j^0| \equiv |\eta_j| > 0$. Let M be so large that $|\eta_j|^2 M > 2R$, with $B(0, R) \supset \Omega$, for all j's. Consider the map:

$$f_d(x_0, z) = \left(\left(x_0 + \sum_1^r(|z_j| - |\eta_j|)^2 M + i(|z_1| - |\eta_1|)\right)^d z_1^{k_1},\right.$$

$$\left. z_j^{k_j}\left((\Pi_1^{j-1}\bar{z}_q^{\alpha_{q,j-1}})^{m_j/l_j} z_j^{l_{j-1}/l_j} - \varepsilon_j\right)_{j=2,\ldots,r}, z_{j+r}^{k_{j+r}}, \ldots\right)$$

where, if $m_1 = |H|$, $\alpha_{q,j-1} = 0$ for $q > 1$ and $\alpha_{1,j-1} = 1$ (in this case $l_j = m_1$ for all j's). $\varepsilon_j \neq 0$ are chosen such that one has the following control on the zeros: $f_d(x_0, z) = 0$ for $z_1 = 0$ and then $z_j = 0$ for all j's (thus outside of Ω), or for $|z_1| = |\eta_1|$, $z_j = 0$ for some j and then $x_0 \leq -M|\eta_j|^2 < -R$ (again outside of Ω), or finally $|z_1| = |\eta_1|, |z_j| = |\eta_j|, x_0 = 0$. This procedure defines ε_j and corresponds to a unique orbit in Ω.

By looking at the index of each zero, as in the proof of Theorem 3.2, it is easy to see that f_d gives a degree equal to $d[F_H]_{S^1}$.

d) If $H = S^1$, that is for d_0, as in [**I.M.V**, Proof of Theorem 4.9 (ii) b], one obtains any suspension in $\Pi_k(S^l)$, via $f(x_0, z) = \left(\|x_0\| f_0(x_0 R/\|x_0\|), z_1^{k_1}, z_2^{k_2}, \ldots\right)$, where $f_0(x_0 R/\|x_0\|)$ represents an element of $\Pi_{k-1}(S^{l-1})$ and assuming that $x_0^0 = 0$ belongs to Ω^{S^1}. Thus the only problematic case would be $l = 2$ which explains our provision.

e) Finally if an element of $\Pi_{k+2m}^{S^1}(S^{l+2n})$ is given by $\sum_0^s d_j [F_{H_j}]_{S^1}$, one may choose the η_j's and x_0^0's, for each subgroup H, such that the zeros are located on different orbits.

More precisely, if $l \geq 1$, let x be the first component of x_0 and perturb slightly this component so that the chosen points in Ω have the following order on x: $x_0 = 0 < x_1 < \ldots < x_s$. Take ε so small that $0 < x_j - x_{j-1} < 4\varepsilon$. Let $\phi_j(x)$ be a smooth Uryshon function with value 1 if $|x - x_j| \leq \varepsilon$ and value 0 if $|x - x_j| \geq 2\varepsilon$. Choose f_0 (giving the invariant degree d_0) such that the first component is positive for $x_0 = (\pm R, y, \tilde{x}_0)$ (this can always be achieved as explained in [**I.M.V**, Appendix A]). In the definition of $f_{(-1)^k d_j, H_j}$ (given in (a) and (b)), replace λ by $x - x_j + iy$.

Define $f(x_0, z)$ as $\phi_j f_{(-1)^k d_j, H_j}(x_0, z) + (1 - \phi_j, 0, \ldots, 0)$ if $|x - x_j| \leq 2\varepsilon, j = 0, \ldots, s$, and as $(1, 0, \ldots, 0)$ otherwise.

If $f(x_0, z) = 0$ then, for some unique j, one has $(x - x_j + iy)^{d_j} z_1^{k_1} = 0$, if $j > 0$, (z_1 may change according to the isotropy subgroup H_j). Then if $z_1 = 0$ the first component is $\phi_j |\eta_1| + (1 - \phi_j)$ or $\phi_j |\eta_1|^\beta \Pi |\eta_r| + (1 - \phi_j)$ which is strictly positive. Thus $x - x_j = 0, \phi_j(x) = 1, \phi_r(x) = 0$ for $r \neq j$, and the zeros of $f(x_0, z)$ correspond to the unique orbit going through $(x_0^0, z_0^1, \ldots)$. For $j = 0$ the same argument, together with the homogeneity of f_0, will give that the zero is for $x = 0$ and then $\phi_0(x) = 1$.

Now on $\Omega \cap \{x/|x - x_j| \leq 2\varepsilon\} \equiv \Omega_j$ one may deform ϕ_j to 1 without moving the zeros and $\deg_{S^1}(f, \Omega_j) = d_j [F_{H_j}]_{S^1}$.

Furthermore $\Sigma_0 \deg_{S^1}(f; \Omega) = \sum_0^s \Sigma_0 deg_{S^1}(f; \Omega_j)$.

From the suspension Theorem 3.3: $d_j(f) = d_j, j \geq 1, d_0(f) \equiv d_0$ modulo 2 (thus the result is proved for $l \neq 2$). Furthermore if $l = 2$, one takes only maps with $d_0 = 0$ and one gets the result (note also that the zeros of f on Ω^{S^1} are those of f_0, thus $d_0(f) = d_0$).

If $l = 0$, then $\Omega^{S^1} = \emptyset$ (and $d_0 = 0$). For the group H with corresponding x_0 and $\phi(x)$, define $f(x, z)$ in the strip $|x - x_0| \leq 2\varepsilon$ by the following modification of the map given in (c):

$$f_d(x, z) = \left((1 - \phi + \phi(x - x_0 + \sum_1^r(\phi|z_j| - |\eta_j|)^2 M) + i(\phi|z_1| - |\eta_1|))^d z_1^{k_1},\right.$$
$$\left. z_j^{k_j}(1 - (\Pi_1^{j-1}(\phi\bar{z}_q)^{\alpha_{q,j-1}})^{m_j/l_j}(\phi z_j)^{l_{j-1}/l_j}\varepsilon_j^{-1})_{j=2,\dots,r}, z_{j+r}^{k_{j+r}}, \dots\right)$$

where $\bar{z}^\alpha = z^{-|\alpha|}$ if $\alpha < 0$. The order of the factors in front of $z_j^{k_j}, j = 1, \dots, r$, may change from group to group.

For the complement of the intervals $|x - x_0| \le 2\varepsilon$, one defines $f(x, z)$ as $(z_1^{k_1}, z_2^{k_2}, \dots, z_m^{k_m}, 0, \dots, 0)$, which is non-zero in Ω since $z = 0$ doesn't belong to Ω.

Again, for $|x - x_0| \le 2\varepsilon$, a zero of f_d would be for $z_1 = 0$ and then $z_j = 0$ for all j's (outside of Ω), or $\phi(x) > 0, \phi|z_1| = |\eta_1|$ and some $z_j = 0$ (but then $|x - x_0| > 2R$ by the choice of M), or $\phi|z_j| = |\eta_j|, j = 1, \dots, r, x - x_0 + (1 - \phi)/\phi = 0$.

Since ϕ is non -increasing, this expression is an increasing function of x with a unique zero in $x = x_0$. Hence $\phi(x_0) = 1$ and one has again a unique orbit in Ω.

The rest of the argument is as in the case $l \ge 1$.

(3) The proof is clear, once one uses Remark 3.3.

Q.E.D.

4.2. Infinite dimensional degree.

Let E be a S^1-Banach space, as in [**I.M.V**, Sections 3 and 4.4], and let Ω be a S^1-invariant open bounded subset of $\mathbb{R}^M \times E$. Let

$$f(x, y) = (f_N(x, y), f_\infty(x, y)) : \Omega \to \mathbb{R}^N \times E$$

be a compact S^1-map such that the S^1-map:

$$\Phi(x, y) = (-f_N(x, y), y - f_\infty(x, y))$$

is non zero on $\partial\Omega$. As in [**I.M.V**, Section 3], one may define the S^1-degree of Φ with respect to Ω as an element of $\Pi^{S^1}_{M,N}$, the stable group of S^1-homotopies (see the above reference for details).

Let $\mathbb{R}^M = \mathbb{R}^k \times \mathbb{C}^m, \mathbb{R}^N = \mathbb{R}^l \times \mathbb{C}^n$ as before, with $k = l + 1 - 2p$, $m = n + p$ if $p \geq 1$. The action of S^1 on $\mathbb{C}^m$ and $\mathbb{C}^n$ is given by the modes m_j and n_j with $n_j = k_j m_j$ for $j = 1, \ldots, n, n_j$ multiples of m_r for all $r = n + 1, \ldots, m$ if $p \geq 1$ and with $n_j = k_j m_j$ only for $j = 1, \ldots, m$ if $p = 0$. For $p = 1$, we shall assume that $n > 0$.

If $p \geq 1$, we shall suppose that whenever H is an isotropy subgroup for the action of S^1 on E, then either $H = S^1$ or, if $|H| < \infty$, then $|H|$ is also a multiple of m_r for all $r = n + 1, \ldots, m$.

Thus if $p \geq 1$, Σ_0 and Σ_1 are isomorphisms (see Theorem 3.3). Furthermore, as in [**I.M.V**, Sections 3 and 4.4], one may approximate the S^1-map Φ by finite dimensional S^1-maps Φ_n with a well defined S^1-degree. One has then:

Proposition 4.1. *If $p \geq 1$, then $\deg_{S^1}(\Phi; \Omega)$ belongs to $\mathbb{Z}$ if $p > 1$ and to $\mathbb{Z}_{(\Pi k_j) m_m / m_0}$ if $k = l - 1$ with $m_0 = (m_1 : \ldots : m_m)$. Moreover, $\deg_{S^1}$ has all the properties of a degree.*

Furthermore $\Pi_{M,N}^{S^1} = \lim_{\longrightarrow} \Pi_{k+2m}^{S^1}(S^{l+2n}) = \begin{cases} \mathbb{Z} & \text{if} & p > 1 \\ \mathbb{Z}_{(\Pi k_j)(m_m / m_0)} & \text{if} & p = 1. \end{cases}$

Proof. It is enough to follow the arguments of [**I.M.V**, Proposition 3.1].

                                                                Q.E.D.

If $k = l + 1$, then from the suspension Theorem 3.3, Σ_0 is one to one except if the dimension of the invariant part is 2, while Σ_1 is always a one to one map (this is the only fact needed for the approximation procedure of [**I.M.V**, Section 3].). One has then:

Proposition 4.2. *If $k = l + 1$, let $l_0 = \dim E^{S^1}$, then*

$$\Pi_{M,N}^{S^1} = \Pi_{k+l_0}(S^{l+l_0}) \times \mathbb{Z} \times \mathbb{Z} \times \ldots$$

with as many copies of $\mathbb{Z}$ as there are isotropy subgroups H for $\mathbb{R}^M \times E$ with $m_H = n_H$ (there may be an infinite number of them). Furthermore

$$\deg_{S^1}(\Phi; \Omega) = \sum d_j [F_{H_j}]_{S^1}$$

is well defined and has all the properties of a degree, provided that if $l + l_0 = 2$, one works with maps which are non-zero on Ω^{S^1} (or one may suspend artificially once by Σ_0). Almost all d_j's are 0.

Proof. It is enough to apply the suspension Theorem 3.3 and the results on the range of the degree, Theorem 4.1. Since $f(x, y)$ is compact, by the approximation procedure, one has that $\deg_{S^1}(\Phi_n; \Omega_n)$ has only a finite number of d_j's which are non-zero and, from the suspension theorem, these numbers don't change for large n.

$$\text{Q.E.D.}$$

Remark 4.1. It is clear that one may extend the S^1-degree for the usual situations of non -compactness taking the necessary technical precautions.

4.3. Computation of the S^1-degree.

For the case $k = l + 1$, the construction given in Theorem 3.2 might be difficult to apply. In this section we shall give a method derived from [**I.M.V**, Remark 4.1] and close to the ideas of [**D.G.J.M**] (although our homotopies are on the range while theirs are on the domain).

In order to reduce the computation to the calculus of extensions degrees we shall use the following construction. Let H be an isotropy subgroup such that $m_H = n_H$.

Decompose V as $V^H \oplus V^\perp$, z as $z^H \oplus z^\perp$ and

$$f(x_0, z) = f^H(x_0, z^H, z^\perp) \oplus f^\perp(x_0, z^H, z^\perp).$$

Since $f^\perp(x_0, z^H, 0) = 0$ and $f(x_0, z) \neq 0$ on $\partial\Omega$, then $f^H(x_0, z^H, 0) \neq 0$ on $\partial\Omega^H$. Thus for ε small enough, one has that $f^H(x_0, z^H, z^\perp) \neq 0$ for $(z^H, z^\perp)$ in a neighborhood of $\partial\Omega$ and $|z^\perp| \leq 2\varepsilon$ (in the infinite dimensional case this follows from a standard compactness argument).

Let $\psi : \mathbb{R}^+ \to [0, 1]$ be a continuous function such that $\psi(r) = 0$ if $r \geq 2\varepsilon$ and $\psi(r) = 1$ if $r \leq \varepsilon$.

The action on $z^\perp \equiv (z_1, z_2, \ldots)$ is given by $(e^{im_1\phi}z_1, e^{im_2\phi}z_2, \ldots)$. Let $n_j = k_j m_j$, and let $z^{\perp k}$ denote the vector $(z_1^{k_1}, z_2^{k_2}, \ldots)$. Then it is easy to see that the homotopy

$$(f^H(x_0, z^H, z^\perp), (1 - \tau\psi(|z^\perp|))f^\perp(x_0, z^H, z^\perp) + \tau\psi(|z^\perp|)z^{\perp k})$$

is valid on $\partial\Omega$ and it is equivariant: since $m_H = n_H$ and $z^{\perp k}$ belongs to the second component in the range. Thus

$$\deg_{S^1}(f;\Omega) = \deg_{S^1}(f_H;\Omega)$$

where

$$f_H(x,z) = (f^H(x,z),(1-\psi(|z^\perp|))f^\perp(x,z) + \psi(|z^\perp|)z^{\perp k}).$$

Since $f_H(x,z) \neq 0$ for $|z^\perp| = \varepsilon$, the S^1-degree is equal to the sum of the degree of f_H on the sets $\Omega \cap \{z^\perp : |z^\perp| < \varepsilon\}$ and $\Omega \cap \{z^\perp : |z^\perp| > \varepsilon\} \equiv \Omega_\perp$. One may replace the first set by $\Omega^H \times \{z^\perp : |z^\perp| < \varepsilon\}$ and deform f_H to $(f^H(x_0, z^H, 0), z^{\perp k})$. Hence

$$\deg_{S^1}(f;\Omega) = \Sigma_k \deg_{S^1}(f^H;\Omega^H) + \deg_{S^1}(f_H;\Omega_\perp)$$

where Σ_k corresponds to the suspension by $k_1, k_2, \ldots$.

Since $m_1, m_2, \ldots$ are not multiples of $|H|$ (because $z_1, z_2, \ldots$ are orthogonal to V^H) and $|H|$ divides the order of any other isotropy subgroup of V^H, the set of indices I_q in Theorem 3.3 is empty.

Similarly if j belongs to I_{qr} then, for the corresponding subgroup H_j of V^H, one has that $|H_j|$ divides $k_i m_i$ but not m_i. But then $k_i m_i$ is also a multiple of $|H|$ and m_i is not a multiple (by construction) of $|H|$. Thus one would have $m_H < n_H$, hence I_{qr} is also empty.

If $\deg_{S^1}(f;\Omega) = \sum d_j [F_{H_j}]_{S^1}$ then, from Remark 3.6,

$$\deg_{S^1}(f^H;\Omega^H) = \sum_{H_j \geq H} d_j [F^H_{H_j}]_{S^1} \quad \text{and}$$

$$\Sigma_k \deg_{S^1}(f^H;\Omega^H) = \sum_{H_j \geq H} d_j [F_{H_j}]_{S^1}$$

On the other hand, one may take $\phi(x_0, z)$ in the definition of $\deg_{S^1}(f_H;\Omega_\perp)$, with value 1 on V^H. From Remark 3.6, one will have that $F_H = (2t + 2\phi - 1, \tilde{f}_H)$ is non-zero in V^H and $\deg_{S^1}(f^H;\Omega_\perp)$ has zero components for the indices j's such that $H_j \geq H$.

Thus:

$$\deg_{S^1}(f_H; \Omega_\perp) = \sum_{H_j \not\geq H} d_j [F_{H_j}]_{S^1}.$$

List then all the isotropy subgroups of V in decreasing order, starting from $H_0 = S^1$ (in the infinite dimensional case reduce first to a finite approximation). For each such subgroup H_j, assume $m_{H_j} = n_{H_j}$ and denote by $z = z^{H_j} \oplus z^{\perp_j}$ the above decomposition, with the corresponding $z^{\perp_j k}$ and $\psi_j = \psi(|z^{\perp_j}|)$.

Let $f_0(x_0, z) = (f^{S^1}, (1 - \psi_0)f^{\perp_0} + \psi_0 z^k)$, then f_0 is homotopic to f and $f_0^{\perp_0}(x_0, z) = z^k$, if $|z| \leq \varepsilon$.

Assume that f_j has been constructed in such a way that f_j is S^1-homotopic to f and $f_j^{\perp_l} = z^{\perp_l k}$, if $|z^{\perp_l}| \leq \varepsilon$ for all $l \leq j$. Define then

$$f_{j+1}(x_0, z) = (f_j^{H_{j+1}}, (1 - \psi_{j+1})f_j^{\perp_{j+1}} + \psi_{j+1} z^{\perp_{j+1} k}).$$

For $|z^{\perp_{j+1}}| \leq \varepsilon$, then $\psi_{j+1} = 1$ and $f_{j+1}^{\perp_{j+1}} = z^{\perp_{j+1} k}$, while if $l \leq j$ and $|z^{\perp_l}| \leq \varepsilon$, then $f_j^{\perp_l} = z^{\perp_l k}$ and the same happens for f_{j+1} (the combination of $1 - \psi_{j+1}$ and ψ_{j+1} gives 1). Clearly f_{j+1} and f are still S^1-homotopic.

Let $\tilde{f}$ be the last of these maps f_j. For each j, let B_ε^j be a ε-neighborhood of V^{H_j} and define

$$\Omega_j = \Omega^{H_j} \setminus \bigcup_{0 \leq k < j} \bar{B}_\varepsilon^k \text{ and } \tilde{\Omega}_j = \Omega_j \times B_\varepsilon^j.$$

Proposition 4.3. *Under the above hypothesis, in particular $k = l + 1$ and $m_{H_j} = n_{H_j}, j \geq 1$, then one has*

$$\deg_{S^1}(f; \Omega) = \sum_j \deg_{S^1}(\tilde{f}^{H_j}; \Omega_j)$$

where $\deg_{S^1}(\tilde{f}^{H_j}; \Omega_j) = \deg_{S^1}(f_l^{H_j}; \Omega_j)$, *for all $l \geq j$, has only one possible non trivial component $d_j[F_{H_j}]_{S^1}$ which is obtained as an extension degree for $\tilde{f}^{H_j}$ or for $f_l^{H_j}$.*

Proof. Note first that, for ε small enough, Ω_j is open and non empty if and only if Ω^{H_j} is non empty. In fact, $\Omega_j = \{(x, z^{H_j}, 0) \in \Omega \colon |z^{H_j}| > \varepsilon, |z^{\perp_1}| > \varepsilon, \ldots, |z^{\perp_{j-1}}| > \varepsilon\}$. If $\Omega^{H_j} \neq \emptyset$ then, as in the proof of Theorem 4.1 and using the fact that Ω^{H_j} is open in V^{H_j}, one knows that Ω^{H_j} has at least one point $(x^0, z^0, 0)$ such that all the components of z^0 are non-zero. Choose ε smaller than the absolute value of these components. Now the point $(x^0, z^0, 0)$, with isotropy subgroup H_j, is in Ω_j since otherwise one would have that, for some $l < j$, $V^{\perp_l} \cap V^{H_j} = \{0\}$, hence $V^{H_j} \subset V^{H_l}$, $H_l < H_j$ which contradicts the chosen order. (Note also that if $\Omega^{H_j} = \emptyset$ then $\Omega^{H_l} = \emptyset$ for all $H_l > H_j$, in particular $\Omega^{S^1} = \emptyset$). Furthermore it is clear that $\tilde{\Omega}_j \cap \tilde{\Omega}_l = \emptyset$, for $j \neq l$.

Since f^{H_j} and $f_l^{H_j}$ are S^1-homotopic on Ω^{H_j} and $f_l^{\perp_p} = z^{\perp_p k}$ if $|z^{\perp_p}| \leq \varepsilon$ and $l \geq p$, then $f_l^{H_j}$ is non-zero on $\partial\Omega_j$ for $l \geq j$ (on $\partial\Omega_j$ at least one of the components of $f_l^{H_j}$ would be of the form $z^{\perp_l k}$ with absolute value ε^k).

Since $V^{H_l} \cap \Omega_j = \emptyset$ for $l < j$, then one has $2t + 2\phi - 1 = 2t + 1$ on V^{H_l} and the only component of $\deg_{S^1}(f_l^{H_j}, \Omega_j)$ is the extension degree corresponding to j (see Theorem 3.2 and Remark 3.6).

For $\tilde{f}$, if (x, z) is a zero of $\tilde{f}$ with isotropy subgroup H_j, then (x, z) belongs to Ω^{H_j} and in fact to Ω_j since, if for some $l < j$, one has $|z^{\perp_l}| \leq \varepsilon$ then $\tilde{f}^{\perp_l} = z^{\perp_l k} = 0$ would imply that (x, z) belongs to $V^{H_j} \cap V^{H_l}$ and the isotropy subgroup of (x, z) would strictly contain H_j. (Note that the points in Ω_j have just one isotropy type, that is their isotropy subgroup is H_j: in fact no point in Ω_j is invariant under a group $H_l > H_j$, since $|z^{\perp_l}| > \varepsilon$).

Thus $\deg_{S^1}(\tilde{f}; \Omega) = \sum_j \deg_{S^1}(\tilde{f}; \tilde{\Omega}_j) = \sum_j \deg_{S^1}(\tilde{f}^{H_j}; \Omega_j)$ using the suspension theorem and the hypothesis $m_{H_j} = n_{H_j}$.

On the other hand, from the construction of f_j and the separation of the components of the degree, one has

$$\deg_{S^1}(f; \Omega) = \deg_{S^1}(f_0^{S^1}; \Omega_0) + \deg_{S^1}(f_0; \Omega_{\perp_0})$$
$$= \sum_{k=0}^{j} \deg_{S^1}(f_k^{H_k}; \Omega_k) + \deg_{S^1}(f_j; \Omega_{\perp_j})$$

following inductively the construction, where $\Omega_{\perp_j} = \Omega \setminus \bigcap_{k \leq j} \bar{B}_\varepsilon^k$.

Moreover, from the construction one has that

$$\deg_{S^1}(f_j; \Omega_{\perp_j}) = \deg_{S^1}(f_j^{H_{j+1}}; \Omega_{j+1}) + \deg_{S^1}(f_{j+1}; \Omega_{\perp_{j+1}}).$$

Since $f_j^{H_{j+1}} = f_{j+1}^{H_{j+1}}$, the induction step is completed.

Taking into account that the last $\Omega_\perp$ is empty, then

$$\deg_{S^1}(f;\Omega) = \sum_j \deg_{S^1}(f_j^{H_j};\Omega_j).$$

Comparing the two sums and recalling that $\deg_{S^1}(\tilde{f}^{H_j};\Omega_j)$ and $\deg_{S^1}(f_j^{H_j};\Omega_j)$ have just one possibly non-trivial component, corresponding to H_j, these two degrees are equal.

Finally, since f_l and f are S^1-homotopic on $\partial\Omega$, then the same argument for f_l would give $\deg_{S^1}(\tilde{f}^{H_j};\Omega_j) = \deg_{S^1}(\tilde{f}_l^{H_j};\Omega_j) = \deg_{S^1}(f_{lj}^{H_j};\Omega_j)$, where f_{lj} is given by the construction process used for f_l. However it is clear that one may choose $f_{lj} = f_l$, for $j \leq l$, by modifying slightly the functions ψ_k.

Q.E.D.

Remark 4.2. Comparison with the degree of [D.G.J.M.]

In [D.G.J.M], the authors define a degree for S^1-maps when $k = l + 1$ and $k_j = 1$ (thus $m_H = n_H$ for all H).

They start from a "generic" situation: assume that all points in Ω have the same isotropy type H and that there is a S^1-map, g, non-zero on $\partial\Omega$, and $g^{-1}(0)$ is contained in the finite disjoint union of open subsets Ω^j of Ω. For each j there is a point a_j in Ω_j and an affine "slice map" $\varphi_j : B^{l+2m_H} \to \Omega$ with $\varphi_j(y) = a_j + A_j y$, where A_j is a linear isomorphism with $\det A_j = 1$ from R^{l+2m_H} into the normal bundle at a_j, to the orbit of a_j, and $\varphi(B^{l+2m_H})$ is contained in a ϵ-neighborhood of a_j, $B_\epsilon(a_j)$. Furthermore it is assumed that Ω^j is contained in the set generated by $\varphi(B^{l+2m_H})$ under the S^1-flow.

The S^1-degree of g, is defined as the integer $\sum_j deg(g \circ \varphi_j; \varphi_j^{-1}(\Omega^j))$, if $H < S^1$, each of these degrees being a Brouwer degree, or the stable suspension of $deg(g^{S^1},\Omega_j)$ if $H = S^1$.

For a S^1-map h from a set Ω, with one isotropy type, the authors show, by a density argument, that h is S^1-homotopic to a map g as above.

Finally, for a general map f and a general open set Ω, the authors construct a map $\tilde{f}$, via a S^1-homotopy on the domain similar to ours in Proposition 4.3, with the property that $\tilde{f}^{\perp_l} = z^{\perp_l}$ if $|z^{\perp_l}| \leq \epsilon$. The zeros of $\tilde{f}$ are in the sets Ω_l defined in Proposition 4.3, with isotropy type H_l. Then the authors define the l'th-component of the S^1-degree as the degree of $\tilde{f}^{H_l}$ on Ω_l.

As usual in this sort of construction, the core of the paper [**D.G.J.M**] is devoted to proving that the final degree is independent of the approximations and to the density of the generic maps. One has then the following result.

Proposition 4.4. *The degree defined in* [**D.G.J.M**] *and the S^1-degre coincide, up to the suspension of the invariant part.*

Proof. It is enough to prove that the l'th-component of $\deg_{S^1}(\tilde{f}^{H_l}; \Omega_l)$ is the degree defined in [**D.G.J.M**]. Since $\tilde{f}^{H_l}$ is S^1-homotopic to a generic map g, the extension degree of $\tilde{f}^{H_l}$ is the sum of the extension degrees of g on each set Ω^j. Then, since $det A_j = 1$, $\deg(g_0\phi_j; \varphi_j^{-1}(\Omega^j)) = deg(g; N)$ where $N = \varphi_j(B^{1+2m_H})$ is an orthogonal cross-section at a_j. If a_j has the component $|z_1| > \epsilon$ (always possible if $H_l < S^1$), then one may choose z_1 real and positive and one may deform N to the cross-section $N_0 \equiv B_\epsilon(a_j) \cap \Omega^j \cap \{Im z_1 = 0, Re z_1 > 0\}$ where each orbit in Ω^j has just one point. For a more explicit argument, see [**I.M.V**, Proposition 4.13].

Since $\varphi_j(B^{l+2m_H})$ is contained in a ϵ-neighborhood of a_j, then this set doesn't intersect the hyperplane $z_1 = 0$.

One may then choose the map $\phi(x, z)$ with value 1 on $\{z_1 = 0\}$ and on V^{H_k} for $k < l$ (since $\Omega^j \subset \Omega_j$ and from the definition of this last set). Thus from Remark 2.4 (ii), the extension degree of g with respect to Ω_j is the Brouwer degree of $(2t + 2\phi(x, z) - 1, \tilde{g}(x, z))$ with respect to the set $\{0 \le t \le 1, |x| \le R, |z| \le R\} \cap V^{H_l}$ divided by $m_1/|H_l|$ (m_1 corresponding to z_1). Since the zeros of this map are $t = 1/2$ and $g(x, z) = 0$ on $\Omega^j \cap \{Im z_1 = 0, 0 \le z_1 < R\}$ then from the product theorem, the extension degree is $(|H_l|/m_1)deg(g; \Omega^j \cap \{Im z_1 = 0, 0 \le z_1 > R\})$, see [**I.M.V**, Remark 2.6 (g)]. Now $e^{im_1\varphi}z_1 = z_1$ if $\varphi = 2k\pi/m_1$ and at least one of the other components of z is moved by such a rotation provided φ is not a multiple of $2\pi/|H_l|$, thus, this happens for $k = 0, 1, \dots m_1/|H_l| - 1$. (See [**I.M.V**, Proof of Proposition 4.13]). Hence, if N_k is the set $e^{i2k\pi/m_1}N_0$, then $\Omega^j \cap \{Im z_1 = 0, 0 \le z_1 < R\}$ is the disjoint union of the N_k's. Using the flow as a homotopy, one gets $deg(g; N_k) = deg(g; N_0)$. Thus the extension degree is $deg(g; N_0)$.

Finally the invariant part of the degree defined in [**D.G.J.M**] is given through a suspension and a compactification at infinity, construction which is equivalent to the embedding in a higher dimensional ball.

Note that one may also prove the coincidence of the two degrees through the computation of indices of isolated orbits, together with another generic argument, as given in the next Chapter.

Q.E.D.

4.4. Global Continuation.

It is clear that once the degree is defined, one may use it to get the global results about existence and dimension given in [**I.M.V$_0$**], since the notion of Γ-epi maps used in that paper is more general than that of S^1-degree. For the reader's convenience we shall give explicitly the global results that one may obtain in the context of a S^1-action.

For continuation problems, suppose one considers the equation

$$\Phi(x, y, \lambda) = 0$$

in an open bounded S^1-invariant subset Ω of $\mathbb{R}^M \times E \times \Lambda$ where Φ is a S^1-equivariant map with values in $\mathbb{R}^N \times E$ and has the compactness properties of section 4.1. S^1 may also act on Λ (if $k \leq l - 1$ the action on Λ has to preserve the hypothesis on the Λ action on $\mathbb{R}^M$).

Assume that for some λ_0 in Λ^{S^1}, $\Phi(x, y, \lambda_0) \neq 0$ on $\partial\Omega_{\lambda_0} = \partial\{(x, y, \lambda) \in \Omega, \lambda = \lambda_0\}$ and $\deg_{S^1}(\Phi(x, y, \lambda_0); \Omega_{\lambda_0}) \neq 0$. Assume also that $\dim \Lambda^{S^1} > 0$.

Theorem 4.2. *There exists a continuum S of solutions of the equation $\Phi(x, y, \lambda) = 0$ with λ in Λ^{S^1}, such that $\bar{S} \cap \partial\Omega \neq \emptyset$ and S/S^1 has dimension at each point at least $dim\Lambda^{S^1}$.*

Proof. Indeed, the map $\Psi(x, y, \lambda) = (\Phi(x, y, \lambda), \lambda - \lambda_0)$ is S^1-equivariant, non-zero on $\partial\Omega$, and

$$\deg_{S^1}(\Psi(x, y; \lambda); \Omega) = \Sigma^\Lambda \deg_{S^1}(\Phi(x, y, \lambda_0); \Omega_{\lambda_0})$$

where Σ^Λ is the suspension given by $\lambda - \lambda_0$.

Since the suspension is a monomorphism, then $\deg_{S^1}(\Psi; \Omega)$ is non trivial and one may apply the argument given in [I.M.V$_0$].

Q.E.D.

It is clear that if $\deg_{S^1}(\Phi^H; \Omega^H_{\lambda_0})$ is non-zero for some isotropy subgroup H, then one will have global continuation in V^H, that is the solutions will have at least H as an isotropy subgroup.

For example if $M = N, k = l$ and $n_j = k_j m_j$, then

$$\deg_{S^1}(\Phi; \Omega_{\lambda_0}) = \deg(\Phi^{S^1}; \Omega^{S^1}_{\lambda_0})$$

and one will obtain a continuum of stationary solutions, a dissapointing result at this stage.

If $k = l + 1, m = n$ and $n_j = k_j m_j$, then, if

$$\deg_{S^1}(\Phi; \Omega_{\lambda_0}) = \sum d_j [F_{Hj}]_{S^1},$$

one has that

$$\deg_{S^1}(\Phi^H; \Omega^H_{\lambda_0}) = \sum_{H_j \geq H} d_j [F^H_{H_j}]_{S^1}.$$

Thus, if d_j is non-zero then one will get the continuum in Ω^H with $H = H_j$.

On the other hand, one may obtain continua of solutions in case one has a symmetry breaking: assume that one has a family of mappings $\Phi(x, y, \lambda)$ defined on a bounded open subset Ω of $\mathbb{R}^M \times E \times \Lambda$ (Ω is not necessarily invariant and Φ is not necessarily equivariant).

Suppose that for some λ_0, $\Phi(x, y, \lambda_0)$ is S^1-equivariant and non-zero and $\partial\Omega_{\lambda_0}$. Then an easy connectedness argument shows that $\Phi(x, y, \lambda_0)$ is non-zero on $(S^1\bar{\Omega}_{\lambda_0})\backslash\Omega_{\lambda_0}$, where $S^1\bar{\Omega}_{\lambda_0}$ is the image of the set $\bar{\Omega}_{\lambda_0}$ under the action of S^1.

Proposition 4.5. *Assume that* $\setminus_*[\deg_{S^1}(\Phi; S^1\Omega_{\lambda_0})] \neq 0$, *then* $\Phi(x, y, \lambda)$ *has a continuum of solutions, which intersects* $\partial\Omega$ *and which has dimension at each point at least* dim Λ.

Proof.

$$\chi_*[\deg_{S^1}(\Phi; S^1\Omega_{\lambda_0})] = \deg(\Phi; S^1\Omega_{\lambda_0})$$
$$= \deg(\Phi; \Omega_{\lambda_0}),$$

where deg is the generalized degree without action and the second equality comes from excision of $(S^1\Omega_{\lambda_0})\backslash\bar{\Omega}_{\lambda_0}$. The rest of the proof is as in Theorem 4.2, when one forgets the action.

                                                                                    Q.E.D.

In particular, if $k = l, m = n, n_j = k_j m_j$ and deg $(\Phi^{S^1}; \Omega_{\lambda_0}^{S^1}) \neq 0$, one obtains a continuum of solutions which may not be stationary. For example if one considers the differential equation:

$$\frac{dX}{dt} = f(X) + \lambda g(t, X)$$

where $g(t)$ is 2π-periodic in t, then for $\lambda = 0$, one has an equivariant problem. Indeed, at the level of Fourier series: for $X = \sum X_n e^{int}$, then the above differential equation becomes

$$in\, X_n - f_n(X) - \lambda g_n(X) = 0 \quad , \quad n \geq 0,$$

as explained in the introduction.

Whenever defined, $\deg_{S^1}(\Phi(x, y, 0); \Omega_0) = \deg(-f(X_0); \Omega_0^{S^1})$ is the degree of the vector field $-f(X_0)$ with respect to the physical set $\Omega_0^{S^1}$. It this degree is non-zero, one will have 2π-periodic solutions of the perturbed equation. This result was proved directly in [**C.M.Z**].

Similarly, if $k = l + 1, m = n, m_j = n_j$, then from Theorem 3.5, if $\deg_{S^1}(\Phi; S^1\Omega_{\lambda_0}) = \sum d_j[F_{H_j}]_{S^1}$ is such that $\sum d_j$ is odd, then one will have the continuum S.

For example, consider the differential equation

$$\frac{dX}{dt} = f(X) + \lambda g(\nu t, X)$$

where $g(\tau, X)$ is 2π-periodic (not necessarily the minimum period), or equivalently with $\tau = \nu t$,

$$\nu\frac{dX}{d\tau} = f(X) + \lambda g(\tau, X).$$

If for some Ω in $H^1(S^1)$, the equation

$$\nu\frac{dX}{dt} = f(X)$$

has no solutions for (ν, x) in $\partial(I \times \Omega)$, with I an interval around some frequency ν_0, then $\deg_{S^1}(\nu\frac{dX}{d\tau} - f(X), I \times \Omega)$ is well defined. If $\sum d_j$ is odd, then one will get periodic solutions for the original perturbed equation. This is called phase locking or entrainment. The special case of a neighborhood of hyperbolic orbit and the corresponding Arnold's tongues will be treated in next chapter.

4.5. Global bifurcation.

Consider the family of S^1-equivariant mappings

$$\Phi(x, y, \lambda) = \Phi(X, \lambda) = 0$$

from $\mathbb{R}^M \times E \times \Lambda$ into $\mathbb{R}^N \times E$, under the assumptions that:

a) Λ has an invariant decomposition $\tilde{\Lambda} \oplus \hat{\Lambda}$ and there are $\lambda_0 = \tilde{\lambda}_0 + \hat{\lambda}_0$ in Λ^{S^1} and X_0 such that $\Phi(X_0, \lambda_0) = 0$.

b) There is a closed invariant set $\mathcal{S}_0$ of solutions of $\Phi(X, \lambda) = 0$, such that

$\alpha)(X_0, \lambda_0) \in \mathcal{S}_0$ (thus the orbit $(S^1 X_0, \lambda_0)$ is in $\mathcal{S}_0$);

$\beta)$ there are small positive numbers, ρ, η, ϵ such that $2\epsilon < \rho$ and:

$\beta_1)$ if $\rho \leq \| \lambda - \lambda_0 \| \leq \rho + \eta, \hat{\lambda} = \hat{\lambda}_0$ and $dist(X, \lambda; \mathcal{S}_0) \leq 2\epsilon$, then $(X, \lambda) \in \mathcal{S}_0$, if $\Phi(X, \lambda) = 0$;

$\beta_2)$ if $\rho \leq dist(X; S^1 X_0) \leq \rho + \eta$ and $\| \lambda - \lambda_0 \| \leq \rho$, then $dist(X, \lambda; \mathcal{S}_0) \geq 2\epsilon$.

Note that $\| \lambda - \lambda_0 \|$, $dist(X, \lambda; \mathcal{S}_0)$ and $dist(X; S^1 X_0)$ are invariant continuous functions. (β_1) says that in the interval $[\rho, \rho + \eta]$ with $\hat{\lambda} = \hat{\lambda}_0, \mathcal{S}_0$ is isolated, while (β_2) says that near $(S^1 X_0, \lambda_0)$ the set $\mathcal{S}_0$ is essentially flat.

Let $\mathcal{S}$ be the set of zeros of $\Phi(X, \lambda)$ and $\tilde{\mathcal{S}}_0 = \mathcal{S}_0 \cap \{\hat{\lambda} = \hat{\lambda}_0\}$.

Since $dist(X, \lambda; \mathcal{S}_0) \leq dist(X, \lambda; S^1 X_0, \lambda_0) \leq max\ dist(X; S^1 X_0), \| \lambda - \lambda_0 \|)$ and $2\epsilon < \rho$, there is some redundancy in this definition. This set is meant to be an invariant neighborhood of $(S^1 X_0, \lambda_0)$ and $\mathcal{S}_0$ may return to λ_0.

Set $U_\epsilon \equiv \{X, \lambda)/\, dist\,(X, \lambda; \mathcal{S}_0) < 2\epsilon, \parallel \lambda - \lambda_0 \parallel < \rho,\, dist\,(X; S^1 X_0) < \rho\}$ and let $\tilde{U}_\epsilon$ be defined as U_ϵ but with λ replaced by $\tilde{\lambda}$ and $\mathcal{S}_0$ by $\tilde{\mathcal{S}}_0$.

Consider the equivariant map defined from $\mathbb{R}^M \times E \times \Lambda$ into $\mathbb{R}^N \times E \times \hat{\Lambda} \times \mathbb{R}$

$$\Psi_\epsilon(X, \lambda) = (\Phi(X, \lambda), \hat{\lambda} - \hat{\lambda}_0,\, dist\,(X, \lambda; \mathcal{S}_0) - \epsilon)$$

and the corresponding restriction $\mathring{\Psi}_\epsilon(X, \tilde{\lambda}) = (\Phi(X, \tilde{\lambda}),\, dist\,(X, \tilde{\lambda}; \tilde{\mathcal{S}}_0) - \epsilon)$ from $\mathbb{R}^M \times E \times \tilde{\Lambda}$ into $\mathbb{R}^N \times E \times \mathbb{R}$.

Since $dist\,(X, \lambda; \mathcal{S}_0) \leq dist\,(X, \lambda; \tilde{\mathcal{S}}_0)$, it is easy to see that, from the above hypotheses, $\Psi_\epsilon(X, \lambda)$ is non zero on ∂U_ϵ and $\mathring{\Psi}_\epsilon(X, \tilde{\lambda})$ is non zero on $\partial \tilde{U}_\epsilon$. Thus, the S^1-degrees of these maps with respect to the corresponding invariant sets are well defined.

Furthermore one may perform a linear deformation from $dist\,(X, \lambda; \mathcal{S}_0) - \epsilon$ to $dist\,(X, \lambda; \tilde{\mathcal{S}}_0) - \epsilon$ on ∂U_ϵ. The zeros of the map in $\{(X, \lambda)/\, dist\,(X; S^1 X_0) < \rho, \parallel \lambda - \lambda_0 \parallel \leq \rho\}$ are in $\tilde{U}_\epsilon \times \{\hat{\lambda}/ \parallel \hat{\lambda} - \hat{\lambda}_0 \parallel < \rho\}$, thus, by excision, one has:

$$\deg_{S^1}(\mathring{\Psi}_\epsilon; U_\epsilon) = \Sigma_{\hat{\Lambda}}\, \deg_{S^1}(\mathring{\Psi}_\epsilon; \tilde{U}_\epsilon).$$

Theorem 4.3. *Let Φ satisfy conditions (a) and (b). Assume that $\deg_{S^1}(\Psi_\epsilon; U_\epsilon) \neq 0$ and let $\mathcal{C}$ be the connected component of (X_0, λ_0) in the closure of $(\mathcal{S}\backslash\mathcal{S}_0) \cup (U_\epsilon \cap \mathcal{S}_0)$, ($\tilde{\mathcal{C}}$ be the corresponding component in $(\tilde{\mathcal{S}}\backslash\tilde{\mathcal{S}}_0)\cup(\tilde{U}_\epsilon \cap \tilde{\mathcal{S}}_0)$). Then $\mathcal{C}$ has a connected invariant subset Σ with the following properties:*

1) $\bar{\Sigma}\cap\{\hat{\lambda} = \hat{\lambda}_0\}$ bifurcates from $\tilde{U}_\epsilon \cap \tilde{\mathcal{S}}_0$ and contains a closed connected invariant subset, which is either unbounded on intersect $\tilde{\mathcal{S}}_0$ outside U_ϵ. $\tilde{\Sigma}/S^1$ has local dimension at least one.

2) If $\Sigma/S^1 = \Sigma_1 \cup \Sigma_2$, with Σ_i proper and closed subsets, then $dim\,(\Sigma_1 \cap \Sigma_2) \geq dim\, \hat{\Lambda}^{S^1}$. In particular Σ/S^1 has dimension at each point at least $dim\, \hat{\Lambda}^{S^1} + 1$.

3) The set of bifurcation orbits $\bar{\Sigma} \cap \mathcal{S}_0$ has a closed connected invariant subset Σ_0 which intersects $\tilde{U}_\epsilon$ and is either unbounded or meets again $\mathcal{S}_0$ outside $\tilde{U}_\epsilon$ and Σ_0/S^1 has dimension at each point at least $dim\, \hat{\Lambda}^{S^1}$.

4) If C is bounded and returns to $\tilde{S}_0$ near a finite number of orbits $(S^1 X_i, \lambda_i)$, λ_i invariant, such that $\tilde{S}_0$ satisfies local hypotheses similar to (a) and (b) on neighborhoods $\tilde{U}_\epsilon^i$, then

$$\Sigma_i \deg_{S^1}(\tilde{\Psi}_\epsilon; \tilde{U}_\epsilon^i) = 0.$$

5) If, for some subgroup H of S^1, one knows that there is no bifurcation near $(S^1 X_0, \lambda_0)$ in $V^H \equiv (\mathbb{R}^M \times E \times \tilde{\Lambda})^H$, then (1), (2), (3) and (4) hold for a subset Σ not in V^H. The return points, if there are any, belong to either V^H or to S_0.

Proof. The argument proceeds as in [I.M.V$_0$, Theorems 3.1. and 4.2] by showing that one has a Γ-epi map on an appropriate set. However here the S^1-degree has the excision property, while the primitive concept of Γ-epi map has only a scaling property which limits the form of S_0.

Recall that if F and G are real Γ-Banach spaces, S and U arbitrary invariant subsets of F, U open, then an equivariant map $g : U \to G$ is said to be **Γ-epi** if and only if:

(i) there is an open bounded and invariant set V such that

$$g^{-1}(0) \cap S \subset V \subset \bar{V} \subset U,$$

(ii) for any map $h : F \to G$, compact equivariant with bounded support contained in U, then the equation $g(x) = h(x)$ has a solution in $S \cap U$.

Here take $U = S = \mathbb{R}^M \times E \times \Lambda \backslash S_0$ and

$$g(X, \lambda) = (\Phi(X, \lambda), \hat{\lambda} - \hat{\lambda}_0, dist\,(X, \lambda; S_0) - \varphi(X, \lambda, \epsilon)),$$

where $\varphi(X, \lambda, \epsilon)$ is a continuous decreasing invariant function with value ϵ if $\| \lambda - \lambda_0 \| \leq \rho$ and $dist\,(X; S^1 X_0) \leq \rho$ and value 0 if $\| \lambda - \lambda_0 \| \geq \rho + \eta$ or $dist\,(X; S^1 X_0) \geq \rho + \eta$. Thus $g(X, \lambda) = \Psi_\epsilon(X, \lambda)$ on U_ϵ. The zeros of g are those (X, λ) in U_ϵ such that $dist\,(X, \lambda; S_0) = \epsilon$ and $(X, \tilde{\lambda})$ in $\tilde{S}_0$ with either $\| \lambda - \lambda_0 \| \geq \rho + \eta$ or $dist\,(X; S^1 X_0) \geq \rho + \eta$. In particular $g^{-1}(0) \cap S \subset U_\epsilon \cap \{\epsilon/2 < dist\,(X, \lambda; S_0) < 2\epsilon\} \equiv V$.

Furthermore if h is as in (ii), let Ω be the bounded open invariant subset:

$$\{(X,\lambda)/h(X,\lambda) \neq 0\}\cup$$

$$\{\epsilon/2 < dist(X,\lambda;S_0) < 2\epsilon, \| \lambda - \lambda_0 \| < \rho + \eta, \ dist(X;S^1X_0) < \rho + \eta\}.$$

Clearly $\bar{\Omega} \subset U$ and, on $\partial\Omega, h(X,\lambda) = 0$. Also $g^{-1}(0) \cap \bar{\Omega} \subset V$. Hence, from the excision property, $\deg_{S^1}(g - h; \Omega) = \deg_{S^1}(g; \Omega) = \deg_{S^1}(\Psi_\epsilon; V_\epsilon) = \deg_{S^1}(\Psi_\epsilon; U_\epsilon)$ and one has a zero of the equation $g(X,\lambda) = h(X,\lambda)$ in $\Omega \subset S$, that is g is S^1-epi on S.

Replacing g by $\tilde{g}$ (without $\hat{\lambda}$ and λ replaced by $\tilde{\lambda}, S_0$ by $\tilde{S}_0$) one gets that $\tilde{g}$ is S^1-epi on $\mathbb{R}^M \times E \times \tilde{\Lambda}\backslash\tilde{S}_0$ (here if $\deg_{S^1}(\Psi_\epsilon; U_\epsilon)$ is non-zero, this will also be the case for $\deg_{S^1}(\tilde{\Psi}_\epsilon; \tilde{U}_\epsilon)$. the converse will depend on the fact that $\Sigma^{\hat{\Lambda}}$ is a monomorphism).

From [**I.M. V$_0$**, Property 1.5], the maps $(\hat{\lambda} - \hat{\lambda}_0,$ dist $(X,\lambda;S_0)) - \varphi)$ and $(dist(X,\tilde{\lambda};\tilde{S}_0) - \varphi)$ are S^1-epi on $S\backslash S_0$ and $\tilde{S}\backslash\tilde{S}_0$ respectively. Furthermore they are bounded and proper on bounded and closed subsets of $S\backslash S_0$ (from the compactness of Φ, S is locally compact). From [**I.M.V$_0$**, Theorem 3.1], one obtains the set Σ satisfying (2) and such that the first map is S^1-epi on it. Then, by [**I.M.V$_0$** Property 1.5], the second map is S^1-epi on $\Sigma \cap \{\hat{\lambda} = \hat{\lambda}_0\}$ and one obtains $\tilde{\Sigma}$ satisfying (1). The fact that $\tilde{\Sigma}$ bifurcates from $\tilde{U}_\epsilon \cap \tilde{S}_0$ comes from the fact that one may vary ϵ, that $\tilde{\Sigma}$ is independent of ϵ and that one must have a point on Σ for which $dist(X,\tilde{\lambda};\tilde{S}_0) = \epsilon$. On the other hand if $\tilde{\Sigma}$ is bounded and does not return to $\tilde{S}_0$, then one may construct a bounded neighborhood V of $\tilde{\Sigma}$ which is, outside $\tilde{U}_\epsilon$, at a positive distance from S_0. Since $(dist(X,\tilde{\lambda};\tilde{S}_0) - \varphi)$ is zero-epi on $\tilde{\Sigma}\cap V$, it must change sign on $\tilde{\Sigma} \cap \partial V$. but this set is contained in $\tilde{S}_0$ and the map is negative.

(3) follows by adapting the arguments of [**I.M.P.V**, Proposition 4.7] to the equivariant setting and checking that they do not depend on the particular form of S_0.

The proof of (4) follows from the construction of an invariant neighborhood of $\tilde{S}$ such that the only solutions of $\Phi(X,\lambda) = 0$ on its boundary are in $\tilde{S}_0$. Thus $\deg_{S^1}((\Phi, dist(X,\tilde{\lambda};\tilde{S}_0) - \epsilon);\Omega)$ is well defined and independent of ϵ. For large ϵ, it is zero, while for ϵ small enough, it is the sum of the local S^1-degrees.

In order to prove (5), replace S_0 by $S_0 \cup V^H$ in the definitions of U_ϵ and Ψ_ϵ obtaining a set V_ϵ and a map χ_ϵ. Then U_ϵ is contained in V_ϵ and the compact set $\tilde{S} \cap (\bar{V}_\epsilon\backslash V_\epsilon)$ corresponds to the points (X,λ) in $\tilde{S}$ with $dist(X,\lambda;S_0 \cup V^H) \leq 2\epsilon \leq dist(X,\lambda;S_0)$. For this set there is an $\epsilon' \leq \epsilon$, such that $dist(X,\lambda;S_0 \cup V^H) \geq 2\epsilon'$: if this is not true, from the compactness, one would have a sequence in the set converging to (X, λ) in $S_0 \cup V^H$. Since $dist(X, \lambda; S_0) \geq 2\epsilon$, then (X, λ) would be in V^H, contradicting the no-bifurcation hypothesis.

One may then replace ϵ by ϵ' in Ψ_ϵ, deform linearly, on ∂U_ϵ, $dist(X, \lambda; S_0) - \epsilon'$ to $dist(X, \lambda; S_0 \cup V^H) - \epsilon'$ and replace U_ϵ by V_ϵ. The argument then follows, as before, for V_ϵ and $\chi_{\epsilon'}$ since $\deg_{S^1}(\Psi_\epsilon; V_\epsilon) = \deg_{S^1}(\chi_\epsilon; V_\epsilon)$.

Q.E.D.

Note that the section $\hat{\lambda} = \hat{\lambda}_0$ may be replaced, as in [**I.M.P.V**, Theorem 4.2], by any other "surface" $\hat{\lambda} = f(\lambda)$ which is close to the previous one near $(S^1 X_0, \lambda_0)$.

In case dim $\hat{\Lambda} = 1$, one may obtain more information.

Corollary 4.1. *Let Φ satisfy hypoteses (a) and (b) with dim $\hat{\Lambda} = 1$. Assume*

$$\deg_{S^1}(\Phi(X, \hat{\lambda}_0 - \rho, \hat{\lambda}_0); \tilde{U}_\epsilon \cap \{\hat{\lambda} = \hat{\lambda}_0 - \rho\}) \neq \deg_{S^1}(\Phi(X, \hat{\lambda}_0 + \rho); \tilde{U}_\epsilon \cap \{\hat{\lambda} = \hat{\lambda}_0 + \rho\})$$

then the conclusions of Theorem 4.3 are valid.

Furthermore, if $S_0 = \{0\} \times \Lambda$, then $\bar{\Sigma} \cap S_0^{S^1}$ has a closed connected subset Σ_0 of local dimension at least dim $\hat{\Lambda}^{S^1}$, which disconnects $S_0^{S^1}$ in $U_\epsilon \cap S_0^{S^1}$. Moreover, for any pair λ_1, λ_2, with λ_1 belonging to the connected component of $\hat{\lambda}_0 - \rho$ in $S_0^{S^1} \backslash \Sigma_0$ and λ_2 to the component of $\hat{\lambda}_0 + \rho$, then either $\bar{\Sigma}$ is unbounded above the segment joining λ_1 to λ_2 or Σ covers λ_1 or λ_2.

Proof. Since dim $\hat{\Lambda} = 1, \hat{\Lambda}$ is contained in Λ^{S^1}. Under the running hypothesis of this chapter, the suspension by $\hat{\lambda} - \hat{\lambda}_0$ gives a one to one map except in the following case: $dim\, E^{S^1} = 0, l = 1$ and $k = 2$, with an invariant suspension (remember that $\hat{\lambda}$ and $dist(X, \lambda; S_0) - \epsilon$ increase the counts on the dimensions of the invariant subspaces by 1). But then $\tilde{\Psi}_\epsilon^{S^1}$ is a map from $\mathbb{R}^3$ into $\mathbb{R}^2$ and either $\tilde{U}_\epsilon \cap V^{S^1} = \emptyset$ or $X_0 \in V^{S^1}$ and this intersection is a ball. In both cases the term d_0 is the suspension of the class of $\tilde{\Psi}^{S^1}$ in $\Pi_2(S^1) = 0$. Thus, in all cases one has that if $\deg_{S^1}(\tilde{\Psi}_\epsilon; \tilde{U}_\epsilon) \neq 0$, then the hypothesis of Theorem 4.2 is met.

Moreover $\tilde{\Psi}_\epsilon$ is deformable to $(\Phi(X, \tilde{\lambda}, \tilde{\lambda}_0), \rho^2 - |\tilde{\lambda} - \tilde{\lambda}_0|^2)$ on the set $\tilde{V}_\epsilon = \{(X, \tilde{\lambda})/$
$dist\,(X, \tilde{\lambda}; \tilde{S}_0) < 2\epsilon, |\tilde{\lambda} - \tilde{\lambda}_0| < \rho + \eta,\ dist\,(X; S^1 X_0) < \rho\}$. Note that $\deg_{S^1}(\tilde{\Psi}_\epsilon; \tilde{U}_\epsilon) =$
$\deg_{S^1}(\tilde{\Psi}_\epsilon; \tilde{V}_\epsilon)$ and that the linear deformation between the two maps is admissible on $\partial \tilde{V}_\epsilon$.
Since the zeros of the second map are for $\tilde{\lambda} = \tilde{\lambda}_0 \pm \rho$ and $(X, \tilde{\lambda})$ on S_0, one has that
$\deg_{S^1}(\tilde{\Psi}_\epsilon; \tilde{V}_\epsilon)$ is the sum of the S^1-degrees with respect to $\tilde{U}_\epsilon \cap \{\tilde{\lambda}_0 \pm \rho - \epsilon < \tilde{\lambda} < \tilde{\lambda}_0 \pm \rho + \epsilon\}$
and with respect to product neighborhoods. From the suspension theorem, one has that
$\deg_{S^1}(\tilde{\Psi}_\epsilon; \tilde{U}_\epsilon) = \Sigma_0\,(\deg_{S^1}(\Phi; \tilde{U}_\epsilon) \cap \{\tilde{\lambda} = \tilde{\lambda}_0 - \rho\}) - \deg_{S^1}(\Phi; \tilde{U}_\epsilon \cap \tilde{\lambda} = \tilde{\lambda}_0 + \rho))$.

The result follows from Theorem 3.3, except if $dim\,(\mathbb{R}^M \times E)^{S^1} = 3, dim\,(\mathbb{R}^N \times E)^{S^1} = 2$. But then in order to have the full additivity property, we have restricted to maps which have $d_0 = 0$ (one may also define d_0 by suspending the invariant part).

The second part follows from the fact that $(\hat{\lambda}^{S^1} - \hat{\lambda}_0,\ dist\,(X, \lambda; S_0) - \varphi)$ is S^1-epi on $\Sigma \cap (\mathbb{R}^M \times E \times \Lambda \backslash S_0)$, [**I.M.V$_0$**, Theorem 3.1], and the induced map on orbits is zero-epi on $(\bar{\Sigma} \cap S_0^{S^1}) \cap (\Lambda^{S^1} \backslash S_0^{S^1})$, see [**I.M.P.V**, Theorem 4.2]. The argument ends with a direct application of [**I.M.P.V**, Proposition 4.8].

$$\text{Q.E.D.}$$

It is clear that if S_0 is not $\{0\} \times \Lambda$ one may obtain information on $\bar{\Sigma} \cap S_0$ with a similar flavor. The results will depend on the topology of S_0 and we leave this task to the reader.

We shall end this chapter by recalling that, in the cases treated in this paper, one has a complete description of the S^1-degree for the cases $k = l, m = n$, and $k = l+1-2p, m = n+p$.

Thus, for $\deg_{S^1}(\Phi(X, \tilde{\lambda}),\ dist\,(X, \tilde{\lambda}; \tilde{S}_0) - \epsilon; U_\epsilon)$, one has results in three different situations (i): $k = l, m = n, n_j = k_j m_j,\ dim\,\tilde{\Lambda} = 1$; (ii): $k = l + 1 - 2p, m = n + p,\ dim\,\tilde{\Lambda} = 1$; (iii): $k = l, m = n,\ dim\,\tilde{\Lambda} = 2, \tilde{\Lambda}$ invariant. For cases (i) and (ii) one may apply Corollary 4.1 while, for case (iii), one has to use Theorem 4.3.

Remark 4.3. Multiplicity results.

One may relate the bifurcation degree, $\deg_{S^1}(\Psi_\epsilon; U_\epsilon)$ to the S^1-degrees of the bifurcated solutions. Assume (a) and (b) hold and suppose for simplicity that $\hat{\Lambda} = 0$ and Λ is invariant.

Assume also that there is no "vertical bifurcation" in the following sense:

c) There are two constants ρ_1 and ϵ_1 with $0 < \rho_1 < \rho, 0 < \epsilon_1 < \epsilon$, such that $\Phi(X, \lambda) \neq 0$, for $dist\,(X, \lambda; S_0) = 2\epsilon$ *and* $\| \lambda - \lambda_0 \| \leq \rho_1$, *and for* $dist\,(X, \lambda; S_0) = \epsilon_1$ *and* $\rho_1 \leq \| \lambda - \lambda_0 \| \leq \rho$.

Note that the second part could be combined with (β_1) and that both follow if we assume that there is an increasing function $g(r)$, with $g(0) = 0$ such that if $dist\,(X, \lambda; S_0) \leq g(\| \lambda - \lambda_0 \|)$ and $\Phi(X, \lambda) = 0$, then X belongs to S_0. In fact, once ρ is fixed, take $2\epsilon \leq g(\rho)$, which implies (β_1). Then if $\Phi(X, \lambda_0) \neq 0$ for $dist\,(X, \lambda; S_0) > 0$, this will be true also for $\| \lambda - \lambda_0 \| \leq \rho$ for ρ small enough and $dist\,(X, \lambda; S_0) = 2\epsilon$. Choose then $\epsilon_1 < g(\rho_1)$.

If $V_\epsilon = \{(X, \lambda)/\,dist\,(X, \lambda; S_0) < 2\epsilon, \| \lambda - \lambda_0 \| < \rho + \eta,\,dist\,(X; S^1 X_0) < \rho\}$,

then $\deg_{S^1}(\Phi(X, \lambda),\,dist\,(X, \lambda; S_0) - \epsilon; V_\epsilon)$ is well defined. Furthermore this degree is equal to the S^1-degree of the map on $W_\epsilon \equiv V_\epsilon \backslash \{(X, \lambda)/\,dist\,(X, \lambda; S_0 \leq \epsilon_1)\}$ (recall $\epsilon_1 < \epsilon$).

On W_ϵ, the deformation $\tau(dist\,(X, \lambda; S_0) - \epsilon) + (1 - \tau)(\| \lambda - \lambda_0 \|^2 - \rho_1^2)$ is valid: if $dist\,(X, \lambda; S_0) = 2\epsilon$ then either $\Phi(X, \lambda) \neq 0$ for $\| \lambda - \lambda_0 \| \leq \rho_1$ or the deformation is positive. If $\| \lambda - \lambda_0 \| = \rho + \eta$ and $\Phi(X, \lambda) = 0$ then (X, λ) is in S_0, that is not in W_ϵ.

Now, by continuity, (c) is also true for $\rho_1 - \eta_1 \leq \| \lambda - \lambda_0 \| \leq \rho_1 + \eta_1$ if η_1 is small enough. One has then to compute $\deg_{S^1}(\Phi(X, \lambda), \| \lambda - \lambda_0 \|^2 - \rho_1^2; W_1)$ where W_1 is the set $\{\epsilon_1 < dist\,(X, \lambda; S_0) < 2\epsilon, \rho_1 - \eta_1 < \| \lambda - \lambda_0 \| < \rho_1 + \eta_1\}$.

Assume that $\Lambda = \mathbb{R} \times \hat{\Lambda}$, with $\lambda \equiv (\tilde{\lambda}, \hat{\lambda})$ is such that:

d) $\Phi(X, \lambda) \neq 0$ if (X, λ) is in $\bar{W}, \rho_1 - \eta_1 \leq \| \hat{\lambda} - \hat{\lambda}_0 \| \leq \rho_1 + \eta_1$ and $|\tilde{\lambda} - \tilde{\lambda}_0| \leq \rho_1 + \eta_1$. Here $\| \lambda - \lambda_0 \| = max\,(\| \tilde{\lambda} - \tilde{\lambda}_0|, \| \hat{\lambda} - \hat{\lambda}_0 \|)$, i.e., the sup-norm.

In other words, the zeros of Φ are for $\rho_1 - \eta_1 < |\tilde{\lambda} - \tilde{\lambda}| < \rho_1 + \eta_1$ and $\| \hat{\lambda} - \hat{\lambda} \| < \rho_1 - \eta_1$. Then, on W_1, one may perform the deformation

$\tau(\| \lambda - \lambda_0 \|^2 - \rho_1^2) + (1 - \tau)(|\tilde{\lambda} - \tilde{\lambda}_0|^2 - \rho_1^2)$. For $\tau = 0$, the zeros are localized at $\tilde{\lambda} - \tilde{\lambda}_0 = \pm \rho_1$. The argument of Corollary 4.1 will give:

Corollary 4.2. *Assume hypotheses (a) - (d). Then the bifurcation degree*

$$\deg_{S^1}(\Phi(X, \lambda),\,dist\,(X, \lambda, S_0) - \epsilon; V_\epsilon) =$$
$$\deg_{S^1}(\Phi(X, \tilde{\lambda}_0 + \rho_1, \hat{\lambda}); W^+) - \deg_{S^1}(\Phi(X, \tilde{\lambda}_0 - \rho_1, \hat{\lambda}); W^-)$$

where $W^{\pm} = \{(X, \hat{\lambda})/\epsilon_1 < dist\,(X, \tilde{\lambda}_0 \pm \rho_1, \hat{\lambda}; \mathcal{S}_0) < 2\epsilon, \parallel \hat{\lambda} - \hat{\lambda}_0 \parallel < \rho_1\}$.

Note that the hypotheses are met if, in a neighborhood of (X_0, λ_0) the relation $\Phi(X, \lambda) = 0$ implies $(X, \lambda) \in \mathcal{S}_0$ holds in the following cases: $\lambda = \lambda_0$, $dist\,(X, \lambda; \mathcal{S}_0) \leq g(\parallel \lambda - \lambda_0 \parallel)$ as above, and $\parallel \hat{\lambda} - \hat{\lambda}_0 \parallel = \rho_1$.

Note also that if, for $\tilde{\lambda} = \tilde{\lambda}_0 \pm \rho_1$, Φ has a finite number of non trivial solutions, then the bifurcation degree will be the sum of the local indices. If the bifurcation index is of the form $(d_0, d_1, d_2, \ldots)$, with d_j in $\mathbb{Z}$ for $j > 0$, and the bifurcated solution have only one of their corresponding indices different from 0 and equal to ± 1, then one has at least $\sum |d_j|$ different bifurcated solutions. This argument applies for Hopf bifurcation of differential equations. We shall come back briefly on this multiplicity result in Chapter VI.

Note finally that, under hypothesis (b), the deformation

$\tau(dist\,(X, \lambda; \mathcal{S}_0) - \epsilon) + (1 - \tau)(\rho^2 - \parallel \lambda - \lambda_0 \parallel^2)$ is always possible. Hence the bifurcation degree is $\deg_{S^1}(\Phi(X, \lambda), \rho^2 - \parallel \lambda - \lambda_0 \parallel^2; V_\epsilon)$. The new map has its zeros only on $\mathcal{S}_0$ and $\parallel \lambda - \lambda_0 \parallel = \rho$. One may reduce V_ϵ to $V_1 = \{(X, \lambda)/\,dist\,(X, \lambda; \mathcal{S}_0) < \epsilon_1, \parallel \lambda - \lambda_0 \parallel < \rho + \eta\}$. Under hypothesis (c), this bifurcation degree is $\deg_{S^1}(\Phi(X, \lambda), \rho_1^2 - \parallel \lambda - \lambda_0 \parallel; \tilde{W}_1)$, with $\tilde{W}_1 = \{(X, \lambda)/\,dist\,(X, \lambda; \mathcal{S}_0) < \epsilon_1, \rho_1 - \eta_1 \leq \parallel \lambda - \lambda_0 \parallel \leq \rho_1 + \eta_1\}$.

If hypothesis (d) holds for $\tilde{W}_1$ (this implies either that $dim\,\Lambda = 1$, the case of Corollary 4.1, or that $\mathcal{S}_0 \cap \{\lambda/\rho_1 - \eta_1 \leq \parallel \hat{\lambda} - \hat{\lambda}_0 \parallel \leq \rho_1 + \eta_1, |\tilde{\lambda} - \tilde{\lambda}_0| \leq \rho_1 + \eta_1\} = \emptyset$) then one has that the bifurcation degree is

$$\deg_{S^1}(\Phi(X, \tilde{\lambda}_0 - \rho_1, \hat{\lambda}); \tilde{W}^-) - \deg_{S^1}(\Phi(X, \tilde{\lambda}_0 + \rho_1, \hat{\lambda}); \tilde{W}^+),$$

where $\tilde{W}^{\pm} = \{(X, \hat{\lambda})/\,dist\,(X, \tilde{\lambda}_0 \pm \rho_1, \hat{\lambda}; \mathcal{S}_0) < \epsilon_1, \parallel \hat{\lambda} - \hat{\lambda}_0 \parallel \leq \rho_1\}$, each of these degrees is the index of the trivial solution.

Now, under hypothesis (c), (d) for W_1 and $\tilde{W}_1$, $\deg_{S^1}(\Phi(X, \lambda); (W_1 \cup \tilde{W}_1) \cap \{\tilde{\lambda}\,fixed\})$ is well defined for $|\tilde{\lambda} - \tilde{\lambda}_0| \leq \rho_1$ and is constant.

Hence

$$\deg_{S^1}(\Phi; \tilde{W}^+) + \deg_{S^1}(\Phi; W^+) =$$
$$\deg_{S^1}(\Phi; \tilde{W}^-) + \deg_{S^1}(\Phi; W^-).$$

Since the bifurcation degree is $\deg_{S^1}(\Phi; \tilde{W}^-) - \deg_{S^1}(\Phi; \tilde{W}^+)$, one recovers the result of Corollary 4.2. This is the classical argument for bifurcation with one parameter. The reason we have not used it here is that, while for the case of $\dim \Lambda = 1$ (hence for bifurcation of stationary solutions and for branching of periodic solutions, see 4.5.1 and 4.5.2 below) the above argument can be followed, it is not valid for the case of $\dim \Lambda > 1$ and Hopf bifurcation: in the later case $\dim S_0 = 2$ and (d) does not hold for $\tilde{W}_1$.

4.5.1. Bifurcation of stationary solutions and symmetry breaking.

If $k = l, m = n, n_j = k_j m_j$, then $\deg_{S^1}(\Phi(X, \tilde{\lambda}_0 \pm \rho, \hat{\lambda}_0); \tilde{U}_\epsilon \cap \{\tilde{\lambda} = \lambda_0 \pm \rho\})$ is given by the ordinary degree of Φ^{S^1}. Furthermore if $(S^1 X_0, \tilde{\lambda}_0 + \rho, \hat{\lambda}_0)$ in S_0 is non-stationary, that is with finite isotropy subgroup, then one of the modes of X_0 will be non-zero. Since the hypothesis of Corollary 4.1 implies that the orbit of X_0 is isolated, then one may choose a small neighborhood of the orbit where this mode will remain non-zero and the complement $2t + 2\phi(X) - 1$ will be positive when restricted to $\tilde{U}_\epsilon^{S^1} \cap \{\tilde{\lambda} = \tilde{\lambda}_0 + \rho\}$. Thus the S^1-index of the orbit will be zero. This implies that Corollary 4.1 can be applied only if S_0 is contained in V^{S^1}.

A change in the ordinary degree of Φ^{S^1} will imply a bifurcation of invariant solutions, result which can be obtained by considering the usual bifurcation problem in V^{S^1}. On the other hand, if one has a symmetry breaking, that is, if $\Phi(X, \lambda)$ is a member, for $\mu = 0$, of a non-equivariant family $\Phi(X, \lambda, \mu)$, then the bifurcation result of Corollary 4.1 is valid for

$$\deg\left(\Phi(X, \lambda, \mu),\, dist\,(X, \lambda, \mu; S_0) - \epsilon; U_\epsilon\right) = (\Pi k_j)\, \deg\left(\Phi^{S^1},\, dist\,(X_0, \lambda, \mu; S_0^{S^1}) - \epsilon; U_\epsilon^{S^1}\right).$$

Thus, if for example $\Phi(0, \lambda, \mu) = 0$ and $S_0 = \{0\} \times \Lambda \times \mathbb{R}$ and $\Phi^{S^1}(X_0, \tilde{\lambda}_0 \pm \rho, 0)$ have different Leray-Schauder indices near $X_0 = 0$, then one will have a $(\dim \Lambda + 1)$- "surface" bifurcating from S_0 at $\tilde{\lambda}_0$.

For $\mu = 0$, this surface will be a surface of stationary solutions, while for $\mu \neq 0$ the bifurcated solutions may be truly non-stationary.

This is the case of the differential equation

$$\frac{dX}{dt} = f(X, \lambda) + \mu g(X, \lambda, t) \quad , \quad \lambda \in \mathbb{R},$$

with $f(0,\lambda) = g(0,\lambda,t) = 0, g$ 2π-periodic in t. Assume that the equation has the linearization $(A(\lambda) + \mu B(\lambda,t))X$ and that $det\, A(\lambda)$ changes sign at $\lambda = 0$. Then Σ will be a 2-dimensional surface with global bifurcation for each section μ equal to any small constant. Σ_0 will correspond to a "curve" where the fundamental matrix $\Psi(\lambda,\mu,2\pi)$ for $A(\lambda) + \mu B(\lambda,t)$ has 1 as Floquet multiplier. At the level of Fourier series, if $B(\lambda,t) = \sum B_n e^{int}$, then the differential equation is equivalent to

$$in X_n - A(\lambda)X_n - \mu(B * X)_n - h_n(\lambda,\mu,X_j,\bar{X}_j) = 0,$$

where $B * X$ is the convolution. If in doesn't belong to the spectrum of $A(0)$, for $n > 0$, (that is there is no possibility of a Hopf bifurcation for the autonomous system), then, locally, one may solve all equations for the modes X_n in function of X_0 and be reduced to a bifurcation equation of the form

$$(A(\lambda) + \mu B_0(\lambda))X_0 + h(\lambda,\mu,X_0) = 0$$

where the bifurcation points will correspond to the set of (λ,μ) such that $A(\lambda) + \mu B_0(\lambda)$ is not invertible. Since $det\, A(\lambda)$ changes sign at 0, this matrix will also have a determinant which changes sign for μ small enough. It is thus clear that one can prove directly a bifurcation result for this symmetry breaking situation but the passage through the S^1-degree and the morphism χ_* shortens considerably the degree computations.

4.5.2. Bifurcation from a familly of orbits and symmetry breaking.

If $k = l + 1$, $n = m$, $n_j = k_j m_j$ (the case $k = l + 1 - 2p$ doesn't offer any difficulty and is left to the reader), then one obtains bifurcation for the equivariant situation as well as for the symmetry breaking case.

Write any element of $\mathbb{R}^k$ as (X_0,ν), X_0 in $\mathbb{R}^l$, and any element of $\mathbb{R}^M \times E$ as (X,ν).

Assume that $\Phi(X,\nu,\lambda)$ has a familly $\mathcal{S}_0$ of orbits $(X(\lambda),\nu(\lambda))$ parametrized by λ near λ_0. For $\lambda = (\tilde{\lambda},\hat{\lambda}_0)$, with $\tilde{\lambda}$ close to $\hat{\lambda}_0 \pm \rho$, the local S^1-index of $(X(\lambda),\nu(\lambda))$ is given by

$$\sum d_j(\lambda)[F_{Hj}]_{S^1}.$$

It is clear that $d_j(\lambda)$ are locally constant. From Corollary 4.1., a change in one of them will imply the global bifurcation result.

Note that if $(X(\lambda), \nu(\lambda))$ are stationary it is unlikely that one would have an isolated solution with a change of $d_j(\lambda)$ for finite H_j. In fact, if $\Phi(X, \nu, \lambda)$ has a linearization $\Phi_{(X,\nu)}(\lambda) \equiv A(\lambda)$, then $A(\lambda)(e^{i\varphi}X, \nu) = e^{i\varphi}A(\lambda)(X, \nu)$, by considering the relation $\Phi(X(\lambda) + e^{i\varphi}X, \nu(\lambda) + \nu) = e^{i\varphi}\Phi(X(\lambda) + X, \nu(\lambda) + \nu)$.

Thus, $A(\lambda)(X, \nu) = (A_0(\lambda)X_0 + \nu B_0(\lambda), \ldots, A_n(\lambda)X_n, \ldots)$.

If $k_j = 1$, then in the generic situation $A_n(\tilde{\lambda}_0 \pm \rho)$ will be invertible complex operators (of the form identity − compact) which are deformable to I. Hence $\Phi(X, \nu, \tilde{\lambda}_0 \pm \rho)$ is deformable to $(\Phi^{S^1}, X_n, \ldots)$ and its S^1-Index is given by the invariant part, (a situation which is not particularly interesting).

One may thus assume that the isotropy subgroup H_0 of $X(\lambda_0)$ is finite. There is then a neighborhood of λ_0 where the isotropy subgroups of $X(\lambda)$ will be contained in H_0 (the modes of $X(\lambda_0)$ which are non-zero will remain non-zero in a neighborhood).

If $\{z_1, z_2, \ldots\}$ are the non-zero components of $X(\tilde{\lambda}_0 - \rho, \hat{\lambda}_0) \equiv X_-$ and H is any isotropy subgroup, then either V^H contains all the components, in which case H is a subgroup of the isotropy subgroup of X_-, or, for some $j, z_j = 0$ on V^H. One may then take the small invariant neighborhood Ω, used for the computation of the S^1-index at X_- such that $z_j \neq 0$ on it, i.e., the corresponding function ϕ will be 1 on V^H. Thus $\deg_{S^1}(\Phi^H, \Omega^H)$ $= \sum_{H_j \geq H} d_j(\lambda_0 - \rho)[F_{H_j}^H]_{S^1} \equiv 0$, by Remark 3.6.

Hence the only $d_j(\lambda_0 - \rho)$'s which are non-zero are those for which H_j is a subgroup of the isotropy subgroup of X_- and hence, for ρ small enough, of H_0 :

$$\text{Index}_{S^1}(\Phi, X(\tilde{\lambda}_0 - \rho, \hat{\lambda}_0)) = \sum_{H_j \leq H_0} d_j(\lambda_0 - \rho)[F_{H_j}]_{S^1}.$$

From Corollary 4.1., Theorem 3.5 and Remark 3.6, one has then:

Proposition 4.6.

a) Under the above hypotheses, if $d_j(\lambda_0 + \rho)$ is different from $d_j(\lambda_0 - \rho)$, then one has a global bifurcation in the sense of Theorem 4.3.

b) If j corresponds to H_j, then this bifurcation takes place in V^{H_j}.

c) If $k_j = 1$, for all $j's$ and $\sum d_j(\lambda_0 \pm \rho)$ have different parities, then, if Φ is part of a non-equivariant family $\Phi(X, \nu, \lambda, \mu)$, for μ small enough, one will also have global bifurcation.

As in Remark 3.8, one may construct a non-equivariant perturbation such that the perturbed system has no solutions while any equivariant perturbation will give a global bifurcation. Consider the coupled system: $((1 - \rho^2 + i\nu\lambda)z_1, (1 - \rho^2 + i\nu\lambda)z_2)$ where $\rho^2 = |z_1|^2 + |z_2|^2, k = 1, k_1 = k_2 = 1, m_1 = m_2 = 1$. The zeros of this map are the stationary point $z_1 = z_2 = 0$ and the periodic set $\rho^2 = 1, \nu = 0$ for $\lambda \neq 0$, with bifurcation at $\lambda = 0$ with any ν. The index of the trivial solution is 0. For the periodic set, take a spherical neighborhood of $\rho^2 = 1$, perform an equivariant rotation to bring the equation into the form $((1 - \rho^2 + i\nu\lambda)^2 z_1, z_2)$. For fixed $\lambda \neq 0$, the S^1-index will be the degree of the map $(1 - z_1^2 + i\nu\lambda)^2 z_1$, for z_1 in a real neighborhood of 1 and ν close to 0. The map $(\nu, z_1) \to (1 - z_1 + i\nu\lambda)^2$ has index equal to 2 (*Sign* λ). Thus, from the change of S^1-index from -2 to 2 one will have global equivariant bifurcation for any equivariant perturbation. On the other hand the non-equivariant perturbation: $((1 - \rho^2 + i\nu\lambda)z_1 + \tau\bar{z}_2, (1 - \rho^2 + i\bar{\nu}\lambda)z_2 - \tau\bar{z}_1)$ has no other solution than $z_1 = z_2 = 0$.

Note that in this example the set S_0 consists of a family of three-dimensional spheres. However it is clear that the argument of Corollary 4.1 goes through. On the other hand the topological "reason" why examples, with $d(\lambda)$ changing from 1 to -1 with just one orbit for each λ, are difficult to construct is that if one reduces the problem to one mode z, then $\Pi_3(S^2)$ is $\mathbb{Z}$ and not $\mathbb{Z}_2$. Thus in this case one would get the same condition (changes of $d(\lambda)$) also in the non-equivariant case. The analytic "reason" will be explained in the next chapter. For a source of examples consult $[I_1]$.

In the next chapter, Proposition 4.6 will be applied to differential equations.

4.5.3. Hopf bifurcation and symmetry breaking.

It remains to consider the case where $dim\ \tilde{\Lambda} = 2, k = l, m = n$ and $n_j = k_j\ m_j$.

If $\tilde{\Lambda} = \{(\mu, \nu)\}$ and if near (μ_0, ν_0) one has a well defined S^1-degree,

$\deg_{S^1}(\tilde{\Psi}_\epsilon,\ dist\ (X, \tilde{\lambda}; \tilde{S}_0)^2 - \epsilon^2; \tilde{U}_\epsilon) = \sum d_j\ [F_{H_j}]_{S^1}$, then if $d_j \neq 0$ for some j, one will be in the position of applying Theorem 4.3.

Taking $\lambda_0 = 0$, one needs that for $\mu^2 + \nu^2 = \rho^2$, the solutions on $\tilde{S}_0$ must be isolated (with respect to solutions not in $\tilde{S}_0$). In particular if X, in $\tilde{S}_0$, has finite isotropy then, as explained in (4.5.1), its local S^1-degree will be 0 and, under perturbation, that solution will likely disappear.

For example the map $(\mu + i\nu)(|z| - 1)z$ has, for $(\mu, \nu) \neq (0, 0)$, the stationary solution $z = 0$ and the periodic solution $|z| = 1$. The solution $z = 0$ has S^1-degree equal to 1 (the degree of $(2t - 1, z)$) while the solution $|z| = 1$ (for $1 - \epsilon < z < 1 + \epsilon, z$ real) has S^1-degree equal to 0. In this case, $dist\ (X, \lambda; \tilde{S}_0) = |\ |z| - 1|$, and the map $((\mu + i\nu)(|z| - 1)z, (|z| - 1)^2 - \epsilon^2)$, for $1 < z < 1 + 2\epsilon$, has the degree of $((\mu + i\nu), z - 1 - \epsilon)$, that is 1: there is global bifurcation from $(\mu, \nu) = (0, 0)$ for this map and for any equivariant small perturbation.

Similarly the map $((\mu + i\nu)(\rho - 1)^2 z_1, (\mu + i\nu)z_2)$, with $\rho^2 = |z_1|^2$, will give $\tilde{S}_0 = \{(z_1, 0)/|z_1| = 1\}$. For fixed $(\mu, \nu) \neq (0, 0)$, the S^1-degree is that of $((\mu + i\nu)^2(\rho - 1)^2 z_1, z_2)$ (through an equivariant rotation); again this degree is 0. While, for $\mu^2 + \nu^2 = \rho^2$, the S^1-degree of the map, with the condition $dist\ (X, \lambda; \tilde{S}_0)^2 - \epsilon^2 = |z_2|^2 + (|z_1| - \rho)^2 - \epsilon^2$, is 2.

On the other hand, the non-equivariant perturbation

$$((\mu + i\nu)(\rho - 1)^2 z_1 + \tau \bar{z}_2, (\mu + i\nu)z_2 - \tau \bar{z}_1)$$

has no other solution (for $\tau \neq 0$) than $z_1 = z_2 = 0$.

If $\tilde{S}_0$ is contained in V^{S^1}, then one may perform a change of variables such that $\tilde{S}_0 = \{X = 0, \mu, \nu\}$. Since $\tilde{S}_0$ is isolated for $\mu^2 + \nu^2 = \rho^2$, one may linearize $\Phi(X, \tilde{\lambda})$ near $\tilde{S}_0$: in this case $dist\ (X, \tilde{\lambda}; \tilde{S}_0) = \parallel X \parallel$ and $\tilde{U}_\epsilon = \{(X, \mu, \nu)/ \parallel X \parallel < 2\epsilon, \mu^2 + \nu^2 < \rho^2\}$. As in the proof of Corollary 4.1, one has $\deg_{S^1}(\tilde{\Phi}, \parallel X \parallel -\epsilon; \tilde{U}_\epsilon) = \deg_{S^1}(\Phi(X, \tilde{\lambda}), \rho^2 - \mu^2 - \nu^2; \tilde{V}_\epsilon)$, with $\tilde{V}_\epsilon = \{X, \mu, \nu)/ \parallel X \parallel < 2\epsilon, \mu^2 + \nu^2 < (\rho + \eta)^2\}$.

If, on $\mu^2 + \nu^2 = \rho^2$, $\Phi(X,\tilde{\lambda}) = A(\mu,\nu)X + h(\mu,\nu,X)$ with $A(\mu,\nu)$ a family of invertible operators and $h(\mu,\nu,X) = o(\|X\|)$, then, from $\Phi(e^{i\varphi}X,\tilde{\lambda}) = e^{i\varphi}\Phi(\mathbf{X},\tilde{\lambda})$, one has that $A(\mu,\nu)e^{i\varphi} = e^{i\varphi}A(\mu,\nu)$ and $h(\mu,\nu,e^{i\varphi}X) = e^{i\varphi}h(\mu,\nu,X)$.

From Schur's Lemma, $A(\mu,\nu)$ will preserve the equivalent representations. The invertibility of $A(\mu,\nu)$ will then imply that $k_j = 1$ and $A(\mu,\nu)$ splits into invertible operators $A_j(\mu,\nu)$ from V_j into itself, where V_j is the space generated by the mode n_j. One has then, for $X = \oplus X_j$, $A(\mu,\nu)X = \oplus A_j(\mu,\nu)X_j$ and

$$\deg_{S^1}(\Phi(X,\tilde{\lambda}),\rho^2 - \mu^2 - \nu^2;\tilde{V}_\epsilon) = \deg_{S^1}(A_j(\mu,\nu)X_j,\rho^2 - \mu^2 - \nu^2;\tilde{V}_\epsilon).$$

For V^{S^1}, d_0 is the class of $(A_0(\mu,\nu)X_0,\rho^2 - \mu^2)$ in $\Pi_{s+1}(S^s)$, where $s = dim\ V_0 = dim\ V^{S^1}$, that is the J-homomorphism image of the class of $A_0(\mu,\nu)$ in $\Pi_1(GL(\mathbb{R}^s))$. See [$\mathbf{I_1}$].

In Chapter VI we shall show that d_j is $\epsilon\nu_j$, where $\epsilon = \pm 1$ and ν_j is the class of $A_j(\mu,\nu)$ in $\Pi_1(GL(\mathbb{C}^{s_j}))$, $s_j = dim\ V_j$, that is the winding number of $\det_\mathbb{C} A_j(\mu,\nu)$ (see Proposition 6.2) and also that d is zero for any other isotropy subgroup H with $|H| \neq n_j$.

One has then:

Proposition 4.7

a) If one of the d_j's is non-zero, then one has a global bifurcation in the sense of Theorem 4.3. In particular if H_j is an isotropy subgroup for which d_j is non-zero, then one has bifurcation in V^{H_j}.

b) If all d_j's are zero and $A_j(\mu,\nu)$ are invertible for $0 < \mu^2 + \lambda^2 \leq \rho^2$, then there is an equivariant map $h(\mu,\nu,X) = o(\|X\|)$ such that $A(\mu,\nu)X + h(\mu,\nu,X) = 0$ implies $X = 0$.

c) If $\sum d_j$ is odd, then one has a global bifurcation for non-equivariant perturbations. If $\sum d_j$ is even, then, as in (b), there is a non-equivariant perturbation of $A(\mu,\nu)X$ such that there is no bifurcation.

Proof. (a) follows from our previous considerations and Theorem 4.3. Note that the bifurcating branch has isotropy at least H_j. There are many examples in [$\mathbf{I_1}$], [$\mathbf{I.M.V_0}$], [$\mathbf{I.M.V}$] where the bifurcation takes place for a larger subgroup.

(b) comes from [$\mathbf{I_1}$, Remark 1.4].

(c) is the consequence of Theorem 3.5 and the fact that the class of $A(\mu, \nu)$, as a family of real operators, is the parity of $\sum d_j$ (if $M+$ dim $E > 3$). The second part follows from [I$_1$]. Note that the oddness of the sum corresponds to the first results on topological Hopf bifurcation given by Alexander and Yorke.

As simple examples of symmetry breaking, one may use the arguments of [I$_1$]. For instance the map $(\lambda z_1, \lambda z_2)$, with $\lambda = \mu + i\nu, m_1 = m_2 = 1$, has S^1-index equal to 2 ($det\ A(\lambda) = \lambda^2$) and the non-equivariant perturbation $(\lambda z_1 + \bar{z}_2, \lambda z_2 - \tau \bar{z}_1)$ has no other solution than $z_1 = z_2 = 0$. The same example with $m_1 = 1, m_2 = 2$, has a S^1-index equal to (1,1) ($A_1(\lambda) = \lambda, A_2(\lambda) = \lambda$).

Equivalent examples for ordinary differential equations will be given in Chapter VI.

CHAPTER FIVE

S^1-INDEX OF AN ISOLATED NON-STATIONARY ORBIT
AND APPLICATIONS.

Although we have computed the possible values of the S^1-degree and given generators for them, the actual calculation of a specific S^1-degree may be complicated. In Chapter IV, Section 3, we have given an algorithm for the computation. One of the first instances where it is useful to apply this algorithm is for the case of an isolated solution. In this chapter we shall treat the case of an isolated orbit and in the next the corresponding problem for families of stationary solutions.

In this chapter the hypotheses will be the same as those of the preceding chapter.

One may discard the case $k = l$, $m = n$, since, as noted in 4.5.1, the S^1-Index of an isolated orbit in zero.

Let $e^{i\varphi}(x^0, z^0)$ be an isolated orbit of non-stationary zeros of the map $\Phi(x, z)$. (In the infinite dimensional case we shall assume that Φ is a finite dimensional approximation). Then the S^1-index, $\text{Index}_{S^1}(\Phi)$, of the orbit is the S^1-degree of Φ with respect to any sufficiently small neighborhood of the orbit. Since the orbit is non-stationary then $z^0 \neq 0$. Denote by $z_1^0, \ldots z_m^0$ the non-zero components of z^0 and let H_0 be the isotropy subgroup of z_0, hence $|H_0| = (m_1 : m_2 : \ldots : m_m)$. By the action of S^1 one may assume that z_1^0 is real and positive. One may choose the neighborhood for the computation of the index so small that it does not intersect the hyperplanes $z_j = 0, j = 1, \ldots, m$, and correspondly choose the Uryshon function $\phi(x, z)$ with value 1 on each of these hyperplanes. Let

$$\tilde{\Phi}(t, x, z) = (2t + 2\phi(x, z) - 1, \Phi(x, z)).$$

Recall that $\tilde{\Phi}$ is defined on the set $I \times D$, with $D = \{(x,z) : \| x \| < R_0, \| z \| < R\}$.

5.1. The case $p \geq 1$.

From Theorem 3.4, if H is the smallest isotropy subgroup, (thus $H < H_0$), then

$$\Theta_H^*(\tilde{\Phi}) = \Pi_1^{m_H}(m_j/|H|)\text{Index}_{S^1}(\Phi)$$

where $\Theta_H^*(\tilde{\Phi})$ is the Brouwer degree of the restriction of the map $\tilde{\Phi}(t, x, Z_1^{m_1/|H|}, \ldots,$ $Z_{m_H}^{m_H/|H|})$ to Z_1 real, $0 < Z_1 < R$ (note that since $\phi(x, 0, z_2, \ldots, z_{m_H}) = 1$, then $\tilde{\Phi}$ is non-zero on $Z_1 = 0$). From the fact that the map $(Z_2, \ldots, Z_{m_H}) \to (Z_2^{m_2/|H|}, \ldots, Z_{m_H}^{m_H/|H|})$ has degree $\Pi_2^{m_H}(m_j/|H|)$, one has that

$$\Theta_H^*(\tilde{\Phi}) = \Pi_2^{m_H}(m_j/|H|)deg(2t + 2\phi(x,z) - 1, \Phi(x,z); I \times D \cap \{z_1 : 0 < z_1 < R\})$$

As in the proof of Proposition 4.3, $\tilde{\Phi}(x,z)$ has $m_1/|H_0|$ zeros on the section $0 < z_1 < R$, all with the same index I_H, which is the index of each of the zeros of $\Phi(x,z)$ on that section. Hence

$$\Theta_H^*(\tilde{\Phi}) = (\Pi_1^{m_H}m_j/|H|)(|H|/|H_0|))I_H.$$

Proposition 5.1. *If $p \geq 1$, then* $\text{Index}_{S^1}(\Phi) = (|H|/|H_0|)I_H$ *(a congruence in* $\mathbf{Z}_{(\Pi k_j)|H|/m_{m_H}}$ *for $p = 1$, thus* $\text{Index}_{S^1}(\Phi) = 0$ *if $k_j = 1$ and $p = 1$).*

Note that this implies that I_H is a multiple of $|H_0|/|H|$. One could expect that I_H should be 0 if $|H| < |H_0|$. However this is not true in general. As an example take the map $f(x,z)$ defined in the proof of Theorem 4.1 (1). This map has an isolated orbit of zeros and a S^1-index equal to d. From $\mathbb{R}^k \times \mathbb{C}^m \times \mathbb{C}$ (with action $e^{iq\varphi}$ on the last component) to $\mathbb{R}^l \times \mathbb{C}^n \times \mathbb{C}$, consider the map $\Phi(x,z,z_{m+1}) = (f(x,z), z_{m+1}^r)$, with r such that qr is a multiple of m_j for $j = m - p + 1, \ldots, m$.

From the Suspension Theorem 3.3, $\text{Index}_{S^1}(\Phi) = (|H|r/|H_0|)\text{Index}_{S^1}(f)$ and $|H|r/|H_0|$ is an integer. Thus $I_H = dr$ is non trivial although H_0 might not be the smallest isotropy subgroup: it is enough to take $m_0 = |H| < l_0 = |H_0|$ in Theorem 3.3.

5.2. The case $p = 0$.

From Theorem 4.1 (2), we have that

$$\text{Index}_{S^1}(\Phi) = \sum d_j [F_{H_j}]_{S^1}.$$

Since $\phi(x, 0) = 1$, then $d_0 = 0$.

If for some isotropy subgroup H, the fixed point space V^H contains the components $z_1, \ldots, z_m$, then $H \leq H_0$. Thus, if $H \not\leq H_0$, $\phi(x, z)|_{V^H} = 1$ and, from Remark 3.6, $d_j = 0$ for all $j's$ such that $H_j \geq H$. In particular

$$\text{Index}_{S^1}(\Phi) = \Sigma_{H_j \leq H_0} d_j [F_{H_j}]_{S^1},$$

and similarly

$$\text{Index}_{S^1}(\Phi^H) = \sum_{H \leq H_j \leq H_0} [F_{H_j}]_{S^1}$$

for any subgroup of H_0.

If $H \leq H_0$, hence $V^H \supset V^{H_0}$, the arguments of Proposition 5.1 and of [**I.M.V**, Proposition 4.13], will give that

$$\Theta_H^*(\tilde{\Phi}) = (0, (\Pi_1^{m_H} m_j / |H|)(|H| / |H_0|) I_H)$$

where I_H is the index of the Poincaré section, $Imz_1 = 0, Rez_1 > 0$, at (x^0, z^0), in V^H. Thus $I_H = 0$ if $m_H < n_H$. (Remark that in [**I.M.V**, Definition 4.11] the generalized Fuller index was defined as $(|H|/|H_0|)I_H$, in order to have a definition independent of the suspensions).

Now, from Theorem 3.4 (2), one has

$$\Theta_H^*(\tilde{\Phi}) = (0, \sum_{H \leq H_j} (d_j |H| / |H_j|)(\Pi_{s \in J_{H_j}} m_s / |H|)(\Pi_{s \in J_H \setminus J_{H_j}} n_s / |H|))$$

Hence, since $d_j = 0$ if $H_j \not\leq H_0$, one gets

$$I_H = \sum_{H \leq H_j \leq H_0} (\Pi_{s \in J_H \setminus J_{H_j}} k_s) \frac{|H_0|}{|H_j|} d_j.$$

If $H_1,\ldots,H_s$ are the subgroups of H_0, labelled in decreasing order, let I_j denote I_{H_j} (and $d_0 = d_{H_0}$), then, as noted in Remark 3.4,

$$\begin{pmatrix} I_0 \\ I_1 \\ \vdots \\ I_s \end{pmatrix} = A \begin{pmatrix} d_0 \\ d_1 \\ \vdots \\ d_s \end{pmatrix}$$

where A is a lower triangular $(s+1)\times(s+1)$ matrix, with $a_{i_0}=1, a_{ij} = \frac{|H_0|}{|H_j|}\Pi_{s\in J_{H_i}\backslash J_{H_j}} k_s$ if $H_i \le H_j$ (thus $i \ge j$).

Hence $d_0 = I_0$ and denoting by $\bar{I}^T = (I_1 - I_0,\ldots,I_s - I_0)$, $D^T = (d_1,\ldots,d_s)$, $\bar{A} = (a_{ij})_{i,j\ge 1}$, one has

$$\bar{I} = \bar{A}D \quad, \quad D = \bar{A}^{-1}\bar{I}.$$

The fact that D must be an integer vector will put strong limitations on the possible indices I_j : for example $d_1 = (|H_1|/|H_0|)(I_1 - I_0)$, etc. These limitations will be considered in Chapter VII.

In order to simplify the discussion, we shall assume from now on that $k_j = 1$ and we shall consider the case (encountered in the next section) where $|I_j| = 1$, although one could play with other cases.

Proposition 5.2. If $k = l+1, k_j = 1, |I_j| = 1$, then $I_j = I_1$ if $|H_0|$ is an even multiple of $|H_j|$ and $I_j = I_0$ if $|H_0|/|H_j|$ is odd. Moreover, $d_0 = I_0, d_j = 0, j \ge 2, d_1 = (I_1 - I_0)/2$ where $I_1 = I_0$ if $|H_0| \ne 2|H_1|$. Compare with [**D.G.J.M**, Theorem 4.1.].

Proof. $d_1 = (|H_1|/|H_0|)(I_1 - I_0)$ and $|H_0|$ is at least $2|H_1|$.

Thus if $I_1 = I_0$, then $d_1 = 0, I_2 - I_0 = (|H_0|/|H_2|)d_2$ and $|H_0|$ is at least $3|H_2|$: thus $d_2 = 0$ since $|I_2 - I_0| \le 2$. Inductively, if one has $d_j = 0$ for $j < j_0$, then $I_{j_0} - I_0 = (|H_0|/|H_{j_0}|)d_{j_0}$ and thus $d_{j_0} = 0$.

Hence if $I_1 = I_0$, then $d_j = 0, j \ge 1$ and $I_j = I_0$.

If $I_1 = -I_0$, then $|H_0| = 2|H_1|$ and $d_1 = (I_1 - I_0)/2 = I_1 = -I_0$. Furthermore $I_2 - I_0 = a_{21}d_1 + (|H_0|/|H_2|)d_2$, with $a_{21} = 2$ if $|H_1|/|H_2|$ is an integer, i.e., if $|H_0|/(2|H_2|)$ is an integer, and $a_{21} = 0$ otherwise. Thus $d_2 = (|H_2|/|H_0|)(I_2 + I_0)$ in the first case and $d_2 = (|H_2|/|H_0|)(I_2 - I_0)$ in the second case. Since $|H_2|/|H_0| \le 1/3$, one has $d_2 = 0$ and the corresponding relation for I_2 and I_1 or I_0.

Let d_{j_0} be the first non-zero d_j, for $j > 2$. Then for $j < j_0$, one has $I_j - I_0 = a_{j1}d_1$, with $a_{j1} = 2$ if $|H_0|/(2|H_j|)$ is an integer and 0 otherwise. The above argument will imply that $I_j = I_1$ if $|H_0|$ is an even multiple of $|H_j|$ *and* $I_j = I_0$ otherwise.

For j_0, on has $I_{j_0} - I_0 = a_{j_0 1}d_1 + (|H_0|/|H_{j_0}|)d_{j_0}$.

Hence $d_{j_0} = (|H_{j_0}|/|H_0|)(I_{j_0} + I_0)$ if $|H_0|/(2|H_{j_0}|)$ is an integer and $d_{j_0} = (|H_{j_0}|/|H_0|)(I_{j_0} - I_0)$ otherwise. In both cases one has $d_{j_0} = 0$. Thus, $d_j = 0$ for $j \geq 2$.

$$\text{Q.E.D.}$$

Remark 5.1. Global bifurcation and period doubling bifurcation.

Assume one has a family $\Phi(x, z, \lambda)$ such that $(x(\lambda), z(\lambda))$ is a curve of orbits of zeros with isotropy subgroup H_0 and isolated for λ close to, but different from, 0. Thus the S^1-index $(0, d_0(\lambda), d_1(\lambda), \dots)$ of $(x(\lambda), z(\lambda))$ is well defined for $\lambda \neq 0$, where $d_j(\lambda)$ corresponds to $H_j \leq H_0$. It is clear that $d_j(\lambda)$ are constants for λ small, λ positive or λ negative.

From the global bifurcation result given in Proposition 4.6, if for some $j, d_j(\lambda)$ changes when λ crosses 0, then one has a branch of orbits of zeros, different from the original curve, which goes to infinity or returns to the original curve at a value of λ different from 0.

In the special case where the Poincaré indices $I_j(\lambda)$ have absolute value 1 then, if $I_0(\lambda) = d_0(\lambda)$ changes sign as λ crosses 0, then one has a *global* branch of zeros bifurcating at $\lambda = 0$, in V^{H_0}. Note that from the change of the Poincaré index one may deduce the *local* existence of the bifurcation but the global result would involve the definition in a coherent way of Poincaré sections for any solution and thus the necessity to perturb globally to a generic situation.

On the other hand one may have no change of sign in $I_0(\lambda)$ but a change in $I_1(\lambda)$, thus in $d_1(\lambda)$ and so $|H_0| = 2|H_1|$. If furthermore the trivial branch is isolated in V^{H_0}, then one will have *global* bifurcation in V^{H_1} of solutions with isotropy subgroup H_1 near the trivial branch. (The remark about the local versus global character of the bifurcation is valid. Globality comes from the fact that we are dealing with a degree theory where the standard arguments apply. It has to be pointed out that away from the bifurcation point there is no control, unless one has further information, on the isotropy type of the bifurcated branch).

For the general situation of indices which may be of absolute value different from 1, if, for some j_0, $d_{j_0}(\lambda)$ changes as λ crosses 0 and there is no bifurcation in V^{H_j} for all j's with

$H_{j_0} < H_j \le H_0$, then one has bifurcation in $V^{H_{j_0}}$ of a branch of solutions with isotropy H_{j_0} near the trivial branch. Put in other words, if $d_{j_0}(\lambda)$ changes as λ crosses 0, then there will be a global bifurcation in $V^{H_{j_0}}$, by considering $\Phi^{H_{j_0}}$ and the relation

$$\text{Index}_{S^1}(\Phi^{H_{j_0}}) = \sum_{H_{j_0} \le H_j \le H_0} d_j [F_{H_j}]_{S^1}.$$

However, unless one knows before hand that there is no bifurcation in V^{H_j}, for $H_j > H_{j_0}$, one cannot be sure that the isotropy type of the bifurcated branch will be H_j (it is at least H_{j_0}). A similar phenomenon occurs in the classical Hopf bifurcation, as explained in next chapter.

5.3. $p = 0$, hyperbolic orbits.

One may wish to think of the isolated orbit as a regular point for the map Φ. In this case, $\Phi^{-1}(0)$ is a one dimensional curve, thus, one needs that $k = l + 1$. Hence we shall distinguish one of the coordinates of $\mathbb{R}^k$ by denoting it ν. Furthermore, one also expects that the linearization of Φ will be invertible when restricted to a transversal section.

We shall now study a case where $|I_j| = 1$ and which is, by standard transversality arguments, generic.

In [**I.M.V**, Definition 4.14] we have introduced the notion of hyperbolicity for C^1-maps $\Phi(X, \nu) = X - f(X, \nu), X = (x, z)$ in a real S^1-Banach space, f is compact and equivariant from $E \times \mathbb{R}$ into E (thus $k = l + 1, k_j = 1$). Let A be the infinitesimal generator of the action. Let X_0 be a non stationary orbit, $X_0 = (x^0, z^0)$, at ν^0, with isotropy subgroup H_0, such that the component $z_1^0 \ne 0$ (real and positive as before). Consider the linear compact operator from $E \times \mathbb{R}$ into itself

$$K(X, \mu) = (f_X(X_0, \nu^0)X + f_\nu(X_0, \nu^0)\mu, \mu + \text{Im} z_1).$$

In [**I.M.V**, Definition 4.14], we studied the kernel of $I - K$ and we gave the following definition:

the orbit is called *hyperbolic* if $I - K$ is invertible, i.e., if

a)$f_\nu(X_0, \nu^0) \notin Range\ (I - f_X(X_0, \nu^0))$ and

b)$dim ker(I - f_X(X_0, \nu^0)) = 1$, generated by AX_0.

The orbit is called *simply hyperbolic* if it is hyperbolic and if the algebraic multiplicity of 1, as eigenvalue of $f_X(X_0, \nu^0)$, is 1.

Then, the index I of the map $\Phi(X, \nu)$ at (X_0, ν^0) on the Poincaré section $(Re z_1 > 0, \quad Im z_1 = 0)$ is the Leray - Schauder index of the map $(X - f(X, \nu), -Im z_1)$ at (X_0, ν^0) and is equal to $\eta(-1)^{\Sigma n_i}$ where n_i is the algebraic multiplicity of $\lambda_i, \lambda_i > 1$, as eigenvalue of $f_X(X_0, \nu^0)$ and η is a sign which depends on a local projection. (See Remark 5.4 below for a more complete argument. Again in [**I.M.V**, Proposition 4.15] we had considered the generalized Fuller index which is equal to $I/|H_0|$).

Now since $\Phi(X, \nu)$ maps $E^H \times \mathbb{R}$ into E^H, one may specialize the definition of hyperbolicity of Φ^H to any subspace E^H which contains X_0. For any subgroup H of H_0, one considers the corresponding operator $K^H = K|_{E^H \times \mathbb{R}}$. (Note that since $f(X_0, \nu)$ is in E^{H_0} so is $f_\nu(X_0, \nu^0)$). We shall thus use the term of $H - (simple) - hyperbolicity$ and the corresponding index is denoted by I_H.

Let $H \le H_0$ and decompose $E^H = E^{H_0} \oplus E^\perp$ (in the infinite dimensional case we shall suppose that this decomposition is a S^1-splitting; this is true for a Hilbert space and an action by isometries, see [**V**, Theorem 2.5.9, equivariant projections]).

Thus X in E^H can be written as $X^0 \oplus X^\perp, f = f^0 \oplus f^\perp$, and f_X has the form

$$f_X = \begin{pmatrix} f_{X^0}^0 & f_{X^\perp}^0 \\ f_{X^0}^\perp & f_{X^\perp}^\perp \end{pmatrix}$$

Proposition 5.3. *The operators $f_{X^0}^\perp$ and $f_{X^\perp}^0$ are 0. The orbit is H-(simply)-hyperbolic if and only if it is H_0-(simply)- hyperbolic and $I - f_{X^\perp}^\perp$ is invertible. If the orbit is H-hyperbolic for all $H \le H_0$, then the S^1-index of the orbit is given by Proposition 5.2.*

Proof. From $f^\perp(X^0, 0) = 0$, one has $f_{X^0}^\perp(X_0) = 0$.

If S_φ denotes the action (real representation) of S^1, then from the equivariance, $f(S_\varphi X) = S_\varphi f(X)$, one obtains

$$S_\varphi f_X(X_0)X = f_X(S_\varphi X_0)S_\varphi X.$$

Thus, for φ in H_0, $S_\varphi X_0 = X_0$, $S_\varphi = I$ on E^{H_0} and, $S_\varphi f_{X^\perp}^\perp = f_{X^\perp}^\perp S_\varphi$, $f_{X^\perp}^0 (S_\varphi - I)X^\perp = 0$, for all $X^\perp$. Now $S_\varphi - I$ is one to one on $E^\perp$ (if not one would have an $X^\perp$ left invariant by H_0, thus in $E^{H_0} \cap E^\perp = \{0\}$). Since the irreducible representations of S_φ on $E^\perp$ are two-dimensional, $S_\varphi - I$ is also onto and $f_{X^0}^\perp(X_0) = 0$. (This is a particular case of Schur's Lemma). The rest of the proposition is then immediate, since $f_\nu(X_0, \nu^0)$ belongs to E^{H_0} and $|I_H| = 1$, for all H.

Q.E.D.

Remark 5.2. Structure of f_X.

The operator $f_{X^\perp}^\perp$ has also a rigid structure, due to the commuting with S_φ for φ in H_0, i.e., for $\varphi = 2\pi k/|H_0|, k = 0, 1, \ldots, |H_0| - 1$.

In fact, if $|H_0| = q|H|$, let $E^\perp$ be decomposed as $E^1 \oplus E^2$, where E^1 corresponds to the modes m_j which are odd multiples of $|H_0|/2$ (thus $E^1 \neq \{0\}$ only if q is even) and E^2 to the other modes for $|H|$.

Thus if $|H_1| = |H_0|/2$, then $E^{H_0} \oplus E^1 = E^{H_1}$. Denote by S the action of S_φ for $\varphi = 2\pi/|H_0|$ (thus for a multiple k, $S_\varphi = S^k$). Then $S|_{E^1} = -I$ and $(S+I)|_{E^2}$ is invertible. Writing $f_{X^\perp}^\perp$ in the form $\begin{pmatrix} A & B \\ C & D \end{pmatrix}$ on $E^1 \oplus E^2$, the commutation with S will give: $B(S+I) = (S+I)C = 0$, $SD = DS$. Hence $B = C = 0$. Now if X in E^2 is such that $(I-\lambda D)^n X = 0$, for a real λ, then $(I - \lambda D)^n SX = S(I - \lambda D)^n X$, that is SX is also a generalized eigenvector unless $SX = \pm X$ (S is an isometry), but then X would be in E^{H_1}. Thus the algebraic multiplicity of any eigenvalue of D is even.

Then if the orbit is H–hyperbolic (thus $I - A$ and $I - D$ are invertible), the contribution to I_H of the operator D will be $(-1)^{\Sigma n_j}$, n_j the algebraic multiplicity of $\lambda_j > 1$, as eigenvalue of D, that is, this contribution is 1. Hence $I_H = I_0$ if q is odd ($A = 0$), and $I_H = I_1$ if q is even.

Note that by pursuing the relation $SD = DS$ for the subgroups H, one may also perturb the operators in such a way that the concept of H-hyperbolicity would be generic.

Remark 5.3. It is convenient at this point to warn the reader about our definition of hyperbolicity and the meaning of our index computations. In fact, under our hypothesis of hyperbolicity, then one may linearize $f(X, \nu)$ in a neighborhood of (X_0, ν_ϵ^0) and obtain:

$$f(X, \nu) = f_X(X_0, \nu^0)(X - X_0) + f_\nu(X_0, \nu^0)(\nu - \nu^0) + g(X, \nu)$$

with $g(X, \nu) = o(\| X - X_0 \| + |\nu - \nu^0|)$, where $f_X(X_0, \nu^0)$ is a compact linear operator. Write $X - X_0 = Y$, $\nu - \nu^0 = \mu$, decompose E as $ker \, (I - f_X) \oplus F$, with $Y = \alpha A X_0 \oplus Z$, and also as $Range \, (I - f_X) \oplus \mathbb{R}$, with Q the projection onto $Range \, (I - f_X)$. Then $X - f(X, \nu) = 0$ is equivalent to:

$$(I - f_X)Z - \eta Q f_\nu - Q g(X, \nu) = 0$$
$$\eta(I - Q)f_\nu + (I - Q)g(X, \nu) = 0.$$

Since $I - f_X$ is invertible as an operator from F into $Range \, (I - f_X)$ and $(I - Q)f_\nu \neq 0$ (hypothesis (a)), the Implicit Function Theorem will give a unique solution $(Z(\alpha), \eta(\alpha))$ and the usual Leray-Schauder degree will give a global continuation.

If f is equivariant, then $Z(\alpha)$ will correspond to $e^{i\varphi}X_0$ and $\eta(\alpha) = 0$ and the global continuation will be the orbit of X_0. In the non-equivariant case the solution can also be parametrized by the amplitude of the pertubation.

Now, as usual with any degree theory, the interest of an index computation in a "generic" situation consists in having results in non-"generic" situations, in having global results at low costs and not to have to linearize around solutions which are not known a priori. Thus, our results on existence of solutions should be seen in this light. (See also our comments in 4.5.2). In particular the reader should convince himself that the considerations which have lead us to Proposition 5.2 are independent of the fact that X_0 was isolated.

As an example, consider the map:

$$\left((1 - \rho^2 + i\nu)z_1, (1 - \rho^2 - i\nu)z_2\right)$$

with $\rho^2 = |z_1|^2 + |z_2|^2$, $m_1 = n_1 = 1$, $m_2 = n_2 = 2$. We are interested in the solution $\rho = 1, \nu = 0$. For $|H_2| = 2$, the equation reduces to $(1 - |z_2|^2 - i\nu)z_2$. It is easy to check that $|z_2| = 1$ is H_2-hyperbolic and has index $d_0 = 1$ (z_2 real and close to 1, the orientation is given by $(Re \, z_2, \nu)$).

For $|H_1| = 1$, the equation is clearly not hyperbolic for $\rho = 1, \nu = 0$. Changing (z_1, z_2) to (Z_1, Z_2^2) and taking the square of the first component, one obtains $(\lambda^2 Z_1^2, \bar\lambda Z_2^2)$ which through a rotation gives $(\lambda Z_1^2, Z_2^2)$. This map has a S^1-index (for ρ close to 1) equal to $-2 = 2I_1$, thus $I_1 = -1$. Hence, from Proposition 5.2, $d_1 = -1$. Note that the non-equivariant perturbation

$$\left((1 - \rho^2 + i\nu)z_1 + \tau z_2, (1 - \rho^2 - i\nu)z_2 - \tau z_1\right)$$

has only $z_1 = z_2 = 0$ as a solution and that the solution $\rho = 1$ is destroyed. Replacing ν by $\nu\lambda$ one will also destroy the bifurcation of periodic orbits at $\lambda = 0$.

In the context of bifurcation, if the family $\Phi(X, \nu, \lambda) = X - f(X, \nu, \lambda)$ has, for $\lambda \neq 0$, a one parameter family of $\{e\}$-hyperbolic solutions $(X_0(\lambda), \nu^0(\lambda))$, then one may linearize as before and obtain:

$$X - f(X, \nu, \lambda) = (A_0 - A(\lambda))Y - f_\nu(\lambda)\eta - g(Y, \eta, \lambda) = 0,$$

where

$$(Y, \lambda) = (X - X_0(\lambda), \lambda), A_0 = I - f_X(X_0(0), \nu^0(0), 0),$$
$$A(\lambda) = f_X(X_0(\lambda), \nu^0(\lambda), \lambda) - f_X(X_0(0), \nu^0(0), 0),$$
$$f_\nu(\lambda) = f_\nu(X_0(\lambda), \nu^0(\lambda), \lambda) \text{ and } g(Y, \eta, \lambda) = o(\| Y \| + |\eta|).$$

Write $E = \ker A_0 \oplus E_2, Y = Y_1 \oplus Y_2$ and let Q be the projection of E onto *Range* A_0, with $E = (I - Q)E \oplus$ *Range* A_0. If K is the pseudo-inverse of A_0 from

Range A_0 onto E_2, then the above equation can be written, for small λ, as (see [I_1]):

$$(A_0 - QA(\lambda))(Y_2 - C(\lambda)(A(\lambda)Y_1 + f_\nu(\lambda)\eta + g(Y, \eta, \lambda))$$
$$\oplus [B(\lambda)Y_1 + G(Y, \eta, \lambda)] + (I - Q)A(\lambda)(Y_2 - C(\lambda)(A(\lambda)Y_1 + f_\nu(\lambda)\eta + g(Y, \eta, \lambda))).$$

Here

$$C(\lambda) = (I - KQA(\lambda))^{-1}KQ$$
$$B(\lambda) = -(I - Q)A(\lambda)(I - KQA(\lambda))^{-1}$$
$$G(Y, \eta, \lambda) = -(I - Q)(I - A(\lambda)KQ)^{-1}(g(Y, \eta, \lambda) + f_\nu(\lambda)\eta).$$

For small λ, $A_0 - QA(\lambda)$ is invertible and the first piece has a unique solution

$$Y_2 = Y_2(Y_1, \eta, \lambda) = KQf_\nu(0)\eta + 0(\| \lambda Y_1 \| + |\lambda\eta|) + o(\| Y_1 \| + |\eta|)$$

The second piece will then give the bifurcation equation:

$$B(\lambda)Y_1 + (I - Q)f_\nu(\lambda)\eta + h(Y_1, \eta, \lambda) = 0,$$

with $h(Y_1, \eta, \lambda) = o(\|\ Y_1\ \| + |\eta|)$, and $B(0) = 0$.

Since one may take $g \equiv 0$ in the above reduction, to say that $X_0(\lambda)$ is $\{e\}$-hyperbolic, for $\lambda \neq 0$, is equivalent to:

(a) for $\lambda \neq 0$, *dim ker* $B(\lambda) = 1$ and *ker* $B(\lambda)$ is generated by $Y_1(\lambda) = PAX_0(\lambda)$, where P is the projection onto *ker* A_0 and A is the infinitesimal generator of the action.

(b) for $\lambda \neq 0$, $B(\lambda)Y_1 + (I - Q)\ f_\nu\ (\lambda) = 0$ has no solution.

Assume that $(I - Q)f_\nu(0) \neq 0$. Let $P(\lambda)$ be the projection onto the space generated by $Y_1(\lambda), (X_0(\lambda)$ is supposed to be continuous), and set $Y_1 = \alpha Y_1(\lambda) \oplus Z$. Similarly let $Q(\lambda)$ be the projection onto $(I - Q)f_\nu(\lambda)$.

Then the bifurcation equation reduces to:

$$Q(\lambda)B(\lambda)Z + \eta(I - Q)f_\nu(\lambda) + Q(\lambda)h(Y_1, \eta, \lambda) = 0$$
$$(I - Q(\lambda))B(\lambda)Z + (I - Q(\lambda))h(Y_1, \eta, \lambda) = 0.$$

The first equation is solvable for $\eta = \eta(Z, \alpha, \lambda)$ and the second gives the system:

$$\tilde{B}(\lambda)Z + \tilde{h}(Z, \alpha, \lambda) = 0$$

with $\tilde{B}(\lambda)$ invertible for $\lambda \neq 0$, $\tilde{h}(Z, \alpha, \lambda) = o(\|\ Z\ \| + |\alpha|)$.

We may then apply the usual bifurcation theory which guarantees a global bifurcation (in λ with α as an extra parameter as in Theorem 4.3 and [I_1]) if *det* $\tilde{B}(\lambda)$ changes sign as λ crosses 0.

Since $\tilde{B}(\lambda)$ is obtained from $B(\lambda)$ through continuous projections, *det* $\tilde{B}(\lambda) =$
η *det* $B(\lambda)$ *Sign* $(I - Q(\lambda))f_\nu(\lambda)$. Similarly, one may play back the sign of *det* $B(\lambda)$ to the spectral behavior of $I - f_X(X_0(\lambda), \nu^0(\lambda), \lambda)$ near $\lambda = 0$ and obtain, through the index computation of Corollary 4.1:

Proposition 5.4. *Under the above hypotheses, let $n(\lambda)$ be the number of eigenvalues of $f_X(X_0(\lambda), \nu^0(\lambda), \lambda)$ which are larger than 1 (counted according to multiplicity). If $n(0_-) - n(0_+)$ is odd, then one has bifurcation. For the equivariant case, one may consider the change in $n_H(\lambda)$ corresponding to $f_X^H(X_0(\lambda), \nu_0(\lambda), \lambda)$, for any subgroup H of H_0, the common isotropy subgroup of $X_0(\lambda)$.*

Note that we could have worked with the Poincaré section $Im\ z_1 = 0$, or added the equation $\eta + Im\ z_1$, taking out the somewhat unnatural parameter α.

In the examples in 4.5.2 and in this remark, one may not use this linearization process, but the S^1-index computations give positive results for those strongly coupled systems.

Remark 5.4. Computation of I_H.

It might be of independent interest to give a characterization of the orientation factor η. Since f_X is a compact operator and $I - f_X$ has a one-dimensional kernel (generated by AX^0), then $I - f_X$ has a one codimensional range. Let Y^* be a generator of the cokernel. Since $f_\nu \notin Range\ (I - f_X)$ then $f_\nu = \alpha Y^* + (I - f_X)Y$ for some Y, $\alpha \neq 0$.

Let k be the algebraic multiplicity of 1 as eigenvalue of f_X and let Y_{k-1} be such that $(I - f_X)^{k-1}Y_{k-1} = AX^0$ (thus if $k = 1$, $Y_0 = AX^0$). Let $Y_{k-1} = \alpha_{k-1}Y^* + (I - f_X)Z$, then:

Proposition 5.5. *If X^0 is H-hyperbolic, then $\alpha_{k-1} \neq 0$ and*

$$I_H = Sign\alpha\, Sign\alpha_{k-1}(-1)^{n_H}$$

where n_H is the number of eigenvalues (counted according to multiplicity) of $f_X|_{E^H}$ which are larger or equal to 1. The product of the two signs doesn't depend on the possible choices for Y^* and Y_{k-1}. If $k = 1$ one may take $Y^* = AX^0$.

Proof. Since $dimker(I - f_X) = 1$ it is clear that $dimker(I - f_X)^\alpha$ is α if $\alpha \leq k$ and k for larger α's. Thus the ascent of $I - f_X$ is k and one has the decomposition:

$$E = ker(I - f_X)^k \oplus Range\,(I - f_X)^k.$$

Let Y_j be defined by $(I - f_X)Y_j = Y_{j-1}$, $j = 1, \ldots, k - 1$ (Y_j is unique if one specifies some orthogonality to Y_0). Then $(I - f_X)^j Y_j = Y_0 = AX^0$ and $Y_j = (I - f_X)^{k-1-j}Y_{k-1}$.

Clearly Y_j is in $Range(I - f_X)$ for $j < k - 1$, while Y_{k-1} is not (if it would be, then the algebraic multiplicity would be $k + 1$ by generating another element Y_k), that is $\alpha_{k-1} \neq 0$. Also α_{k-1} is independent of the possible choices for Y_{k-1}. (The assertion about the product of the signs is clear). Let $\eta = Sign\alpha\, Sign\alpha_{k-1}$.

Now one has to compute the Leray-Schauder index, at 0, of the map $((I - f_X)X - \mu f_\nu, -Imz_1)$ with respect to the variables (X, μ), where z_1, in $\mathbb{C}$, is such that $(AX^0)_1 = iz_1^0$, with z_1^0 real and positive. Now the homotopy $\tau f_\nu + (1 - \tau)\eta Y_{k-1}$ is valid, since this element has a projection on Y^* which has the sign of $Sign\alpha$, hence it doesn't belong to $Range(I - f_X)$. (Thus a zero will imply $\mu = 0$, X is a multiple of AX^0, with $Imz_1 = 0$, that is $X = 0$).

As explained in [**I.M.V**, Proposition 4.15], one may choose $\lambda < 1$, close to 1, such that the operator

$$(I - \lambda K)(X, \mu) = ((I - \lambda f_X)X - \lambda \eta Y_{k-1}, (1 - \lambda)\mu - \lambda Imz_1)$$

is invertible, as well as $I - \lambda f_X$. Let $Z(\lambda)$ be such that $Y_{k-1} = (I - \lambda f_X)Z(\lambda)$ and let $\alpha(\lambda) = 1 - \lambda - \lambda \eta Imz_1(\lambda)$, where $z_1(\lambda)$ is the corresponding component of $Z(\lambda)$. Then $(I - \lambda K)(X, \mu)$ can be deformed to $((I - \lambda f_X)X, \mu\alpha(\lambda))$ $\quad (\alpha(\lambda) \neq 0$ for λ close to 1), whose index is $Sign\alpha(\lambda)\, Index\, (I - \lambda f_X)$. Since $Index(I - \lambda f_X)$ is $(-1)^{n_H - k}$ if $\lambda < 1$, it remains to prove that $Sign\, \alpha(\lambda) = (-1)^k \eta$.

We claim that $Z(\lambda) = (1 - \lambda)^{-1} \sum_0^{k-1} (-\lambda/(1 - \lambda))^{k-1-j} Y_j$.

In fact, since $Y_j = (I - f_X)^{k-1-j} Y_{k-1}$, one may write the sum as $\sum_0^{k-1} A^j Y_{k-1}$, with $A = (-\lambda/(1 - \lambda))(I - f_X)$ and $I - A = (I - \lambda f_X)/(1 - \lambda)$. But $(I - A) \sum_0^{k-1} A^j Y_{k-1} = (I - A^k)Y_{k-1} = Y_{k-1}$ since $A^k Y_{k-1} = 0$. Since $Z(\lambda) = (I - A)^{-1} Y_{k-1}/(1 - \lambda)$, the claim is proved.

Then $Sign\alpha(\lambda) = Sign(1 - \lambda)^k \alpha(\lambda) = Sign\,[(1 - \lambda)^k - \lambda \eta Im\,(\sum_0^{k-1}(-\lambda)^{k-1-j}(1 - \lambda)^j Y_j)_1]$, where the subscript 1 corresponds to the z_1 component. Since $Sign\, \alpha(\lambda)$ is constant for $\lambda < 1$, close to 1, one may take λ to 1 and obtain

$Sign\, \alpha(\lambda) = -\eta(-1)^{k-1}\, Sign\, ImY_{01} = \eta(-1)^k.$

Q.E.D.

5.4. Autonomous differential equations.

Suppose that

$$\frac{dX}{dt} = g(X), \quad X \in \mathbb{R}^M,$$

has a periodic solution X_0 of period $2\pi/\nu$, (not necessarily minimal), or equivalently that the equation $\nu\frac{dX}{d\tau} = g(X)$ has a 2π-periodic solution $X_0(\tau/\nu)$. Then, if $B(\tau) = Dg(X(\tau/\nu))$, we have seen in [**I.M.V**, 4.5] that the orbit $X_0(\tau)$ is hyperbolic (and in fact simply hyperbolic) if and only if $\dot{X}_0(\tau)$ is a simple eigenvector of $\frac{d}{d\tau} - \frac{1}{\nu}B$. In other terms, let $\Phi(\tau)$ be the fundamental matrix for this linear equation, then $X_0(\tau)$ is hyperbolic if and only if 1 is a simple eigenvalue of $\Phi(2\pi)$.

Thus if $\Psi(t) = \Phi(\nu t)$, then $\frac{d}{dt}\Psi(t) = B(t)\Psi(t)$ and $\Psi(2\pi/\nu)$ is a Poincaré return map.

If $T = 2\pi/\nu$ is a multiple $k_0 T_0$ of the minimal period $T_0 = 2\pi/\nu_0$, then $\Psi(T) = \Psi(T_0)^k$ and the orbit $(X_0(\tau), \nu)$ has $H_0 = \mathbb{Z}_{k_0}$ as isotropy subgroup (the Fourier coefficients of $X_0(\tau)$ are non-zero only for multiples of k_0).

Furthermore, in [**I.M.V**, Proposition 4.16], we have proved that if 1 is a simple eigenvalue of $\Psi(T)$, then for $H = \{e\}$, I_H at (X_0, ν), is $(-1)^{\Sigma n_i}$ where n_i is the algebraic multiplicity of the eigenvalue $\lambda_i > 1$ of $\Psi(T)$, the k_0'th Poincaré return map.

Thus the orbit (X_0, ν) is $\{e\}$-hyperbolic if and only if the Poincaré return map $\Psi(T_0)$ has 1 as a simple eigenvalue and no other eigenvalues which are k_0'th roots of unity.

For a subgroup $H = \mathbb{Z}_k$ of H_0, that is if k divides k_0, then similarly, I_H, at (X_0, ν), will be $(-1)^{\Sigma n_i}$ with n_i corresponding to the map $\Psi(T/k)$.

Hence the orbit will be H-hyperbolic if $\Psi(T_0)$ has no Floquet multipliers which are (k_0/k)-roots of unity, for all divisors of $k_0 = \nu_0/\nu$.

One has then the classical definition of hyperbolicity if one wishes this property for all multiples of T_0.

Since $\Psi(T/k) = \Psi(T_0)^{k_0/k}$, the matrix A, in Remark 5.2, will have 1 as eigenvalue if and only if -1 is a Floquet multiplier for $\Psi(T_0)$ and k_0 is even. Furthermore since the complex roots of unity come by pairs, the parity will not be altered by them.

Let σ_+ be the number of Floquet multipliers for $\Psi(T_0)$ (counted according to the algebraic multiplicity) which are real and larger than 1, σ_- those which are real and less than -1. Then we have the following result.

Proposition 5.6. *If $\Psi(T_0)$ has no (non-trivial) Floquet multipliers which are k_0'th roots of unity, then the index I_H at $(X_0, \nu = \nu_0/k_0)$ for $|H| = k$ a divisor of k_0, is $I_0 = (-1)^{\sigma_+}$ if k_0/k is odd, and $I_1 = (-1)^{\sigma_+ + \sigma_-}$ if k_0/k is even. Furthermore the S^1-index at (X_0, ν) has only two possible non zero components: $d_0 = I_0$ for H_0 and $d_1 = (I_1 - I_0)/2$ for $H_1 = \mathbb{Z}_{k_0/2}$ (thus k_0 must be even in this case).*

Remark 5.5. Fuller's degree.

As seen in Theorem 3.4 and [**I.M.V**, Proposition 4.16], the Fuller degree of the differential equation at ν_0/k_0 is $\sum d_j/|H_j|$. For a hyperbolic orbit, $I_0/k_0 + [(I_1 - I_0)/2]/(k_0/2) = I_1/k_0$ if k_0 is even and I_0/k_0 if k_0 is odd. Thus, in both cases the Fuller index is $I_{\{e\}}/k_0$, recovering the classical result.

Remark 5.6. Bifurcation.

From the above proposition, it is clear that if one has a family of hyperbolic orbits, such that, for some parameter value, one of the multipliers crosses through 1, then one has a bifurcation of periodic orbits. From Remark 5.1, this local bifurcation extends to a global bifurcation from $\nu = \nu_0/k_0$ in V^{H_0}. Similarly, if for some other value of the parameter, one of the multipliers crosses through -1, then one has a local bifurcation of orbits with double period.

In the first case $d_0(\lambda)$ changes, while $d_1(\lambda)$ will change only if σ_- is odd. In the second case, $d_0(\lambda)$ will not change and $d_1(\lambda)$ will go from 0 to ± 1 or conversely. In that case, Theorem 4.3 will guarantee a global bifurcation, from $\nu = \nu_0/k_0$ with k_0 even, in V^{H_1}, the space of Fourier coefficients which correspond to modes which are multiples of $k_0/2$.

Note that in this case the term period doubling used in Remark 5.1 is completely justified.

5.5. Gradient maps.

In [**Da**], Dancer has considered fixed points of maps which are gradient of real valued functions which are S^1-invariant. For such a gradient $f(X)$ we have seen that Dancer's degree and the normalized S^1-degree defined in [**I.M.V**] coincide. Furthermore, for a hyperbolic isolated non-stationary orbit X_0 with isotropy subgroup H_0, that is if $dim\ ker(I - f_X(X_0)) = 1$, then $I_H = (-1)^{\Sigma n_i}$, where $\sum n_i = n_H$ is the number of eigenvalues of $f_X|E^H$ greater than 1 for any H subgroup of H_0 (here f_X, being a Hessian, is symmetric and $\nu^0 = 0$). See [**I.M.V**, Propositions 4.17 and 4.18]. From Propositions 5.2 and 5.3, one has:

Proposition 5.7. *Assume $dim\ ker(I - f_X(X_0)) = 1$ and let $|H_0| = q|H|$. Then $I_H = I_0$ if q is odd and $I_H = I_1$ if q is even. Let $n_{H_1} = n_{H_0} + n_A$ (A given in Remark 4.4), then there is a possibility of period doubling if n_A changes parity. Index$_{S^1}$ (f) at X_0 has only two possibly non-trivial components, $d_{H_0} = I_0$ and $d_{H_1} = (I_1 - I_0)/2$, if $|H_0| = 2|H_1|$.*

5.6. Differential equations with fixed period.

In [**I.M.V**, Proposition 4.19] we have also considered the case of 2π-periodic solutions of the equation $\dot{Y} = g(Y, \nu)$. An instance of such an equation would the autonomous situation.

Let $(X_0(t), \nu_0)$ be a non-stationary periodic solution, with minimal period $2\pi/n_0$. Set $B(t) = Dg(X_0(t), \nu_0)$. Assume that if one has $\dot{Y} - B(t)Y = 0$, then $Y(t) = \alpha\dot{X}_0(t)$ and suppose $\dot{Y} - B(t)Y = g_\nu(X_0, \nu^0)$ has no 2π-periodic solution. Thus, the orbit is $\{e\}$-hyperbolic.

Denote by z_1 a non-zero mode of X_0, as in the previous sections. Let $X(\lambda, t)$ be the unique 2π-periodic solution of $\dot{Y} - (B - \lambda)Y = g_\nu$, for λ small and positive. Let $z_1(\lambda)$ be the corresponding mode of $X(\lambda, t)$. We have shown in [**I.M.V**, Proposition 4.19] that $I_H = -\ Sign\ Im\ z_1(\lambda)(-1)^{n_H}$, where n_H is the number of the Floquet multipliers (counted according to their multiplicity) of the $n_0/|H|$ 'th Poincaré return map. Thus, as before in Proposition 5.5, $(-1)^{n_H} = (-1)^{\sigma_+}$ if $n_0/|H|$ is odd and $(-1)^{n_H} = (-1)^{\sigma_+ + \sigma_-}$ if $n_0/|H|$ is even. Also the orbit will be H-hyperbolic if no (non-trivial) Floquet multiplier of the first return map is a $n_0/|H|$ root of unity.

We shall give a formula for I_H which doesn't depend on the solution $X(\lambda, t)$.

Let $\Phi(t)$ be the fundamental matrix for $B(t)$ and let Z_0 be the generator of $ker(I - \Phi(2\pi)^T)$ (recall that $ker(I - \Phi(2\pi))$ is generated by $\dot{X}_0(0)$ and is one dimensional).

Let $Z(t) = \Phi^{-1T}(t)Z_0$, then $Z(t)$ generates the kernel of $\left(\frac{d}{dt} + B^T\right)$ in $C^1(S^1)^N$ (it is easy to see that Φ^{-1T} is the fundamental matrix for the adjoint problem).

Let k be the algebraic multiplicity of 1 as eigenvalue of $\Phi(2\pi)$ and let e_{k-1} be a vector in $\mathbb{R}^N$ such that $(I - \Phi(2\pi))^{k-1}e_{k-1} = \dot{X}_0(0) = e_0$.

Then, as in Proposition 5.5, one has

Proposition 5.8. *Under the above hypotheses, one has that*

$$I_H = Sign(e_{k-1}.Z_0)Sign(\int_0^{2\pi} g_\nu(X_0(t),\nu^0) \cdot \Phi^{-1T}(t)Z_0 dt)(-1)^k(-1)^{n_H}.$$

Proof. The argument is parallel to the one used in Proposition 5.5. Let e_j be orthogonal to e_0 such that $(I - \Phi(2\pi))e_j = e_{j-1}$, $j = 1, \ldots, k-1$.

Thus $e_j = (I-\Phi(2\pi))^{k-1-j}e_{k-1}$ and $e_j \cdot Z_0 = 0$ for $j = 0, \ldots, k-2$, while $e_{k-1}\cdot Z_0 \neq 0$. (If not the algebraic multiplicity would be more than k). In order to compute the index one may replace g_ν by any $Z_1(t)$ which has $(Z_1(t), Z(t))_{L^2}$ of the same sign as $(g_\nu, Z(t))_{L^2}$.

If η is the product of the two signs in Proposition 5.7, define

$$Z_1(t) = \eta\Phi(t)(\sum_0^{k-1}(t/2\pi)^{k-1-j}e_j).$$

Then $\int_0^{2\pi} Z_1(t) \cdot Z(t)dt = 2\pi\eta(e_{k-1} \cdot Z_0)$ has the right sign.

Since $\Phi(2\pi)\sum_0^{k-1} e_j = \Phi(2\pi)\sum_0^{k-1}(I-\Phi(2\pi))^{k-1-j}e_{k-1} = (I-(I-\Phi(2\pi))^k)e_{k-1} = e_{k-1}$, one has that $Z_1(0) = Z_1(2\pi)$.

Furthermore the 2π-periodic solution of $(\frac{d}{dt} - (B - \lambda))X = Z_1$ is then

$$X_\lambda(t) = \eta e^{-\lambda t}\Phi(t)\left[e^{-2\pi\lambda}\int_0^{2\pi} e^{\lambda s}(I - e^{-2\pi\lambda}\Phi(2\pi))^{-1}\sum_0^{k-1}e_j(s/2\pi)^{k-1-j}ds\right.$$

$$\left. + \int_0^t e^{\lambda s}\sum_0^{k-1}e_j(s/2\pi)^{k-1-j}ds\right]$$

As in Proposition 5.5, one may see that

$$(I - e^{-2\pi\lambda}\Phi(2\pi))^{-1}e_j = (1 - e^{-2\pi\lambda})^{-1}\sum_0^j(-e^{-2\pi\lambda}/(1 - e^{-2\pi\lambda}))^{j-s}e_s$$

(express e_j and e_s in terms of e_{k-1} and use the geometric series argument). Hence by multiplying $X_\lambda(t)$ by $(1 - e^{-2\pi\lambda})^k$ and taking the limit as λ goes to 0 positively,

$$\lim_{\lambda \to 0^+} (1 - e^{-2\pi\lambda})^k X_\lambda(t) = \eta\Phi(t) \int_0^{2\pi} (-1)^{k-1} e_0(s/2\pi)^{k-1} ds = \eta(2\pi/k)(-1)^{k-1}\dot{X}_0(t)$$

Q.E.D.

5.7. Differential equations with first integrals.

We shall use the index computation given in Proposition 5.7 to relate our normalized degree of [**I.M.V**, Proposition 4.19] to a new degree introduced by **Dancer** and **Toland** in [**D.T**] for systems with a first integral. Assume that the equation

$$\dot{X} = g(X) \quad , \quad X \text{ in } \mathbb{R}^N, \text{ has a first integral } V(X).$$

This means that $V(X(t)) = V(X(0))$ on solutions of the equation, or equivalently that $\nabla V(X) \cdot \dot{X} = \nabla V(X) \cdot g(X) = 0$.

Let ω be an open bounded subset of $\mathbb{R}^N$, such that

(a) $g(X) \neq 0$ on $\bar{\omega}$,

(b) $\nabla V(X) \neq 0$ on ω,

(c) $\dot{X} = g(X)$ has no 2π-periodic solutions on $\partial\omega$.

Then Dancer and Toland have defined a rational-valued degree which is built from the generic situation of isolated orbits. This is done in the following way: $X_0(t)$ is an isolated orbit, of minimal period $2\pi/n_0$. Let Q be the projection of $\mathbb{R}^N$ onto the orthogonal complement of $\nabla V(X_0(0))$. Let $F(X,t)$ be the non-linear flow, i.e., the solution of the equation with initial data X, and let E_0 be the affine plane containing $X_0(0)$ and orthogonal to $g(X_0(0))$.

The index of the orbit is defined as the Brouwer index, with respect to E_0, of the map $Q(I - F(X, 2\pi))$ divided by n_0.

In the generic case of a one dimensional kernel for $I - DF(X, 2\pi)$ and an algebraic multiplicity k, then $k \geq 2$ and the Brouwer index is $Sign\, (\nabla V(X_0(0)) \cdot e_{k-1})(-1)^{k+n_H}$.

Now, denote by $B(t)$ the matrix $Dg(X_0(t))$ and let $\Phi(t)$ be the fundamental matrix for the variational equation $\dot{X} - B(t)X$. It is clear, by linearizing $F(X_0(0), t)$, that $\Phi(t) = DF(X_0(0), t)$. Also, by linearizing the identity $V(F(X, t)) = V(X)$, one has that

$$\nabla V(F(X_0, t)) \cdot \Phi(t)W = \nabla V(X_0) \cdot W \quad \text{for all W in } \mathbb{R}^N.$$

Thus, since $\Phi(t)$ is invertible, if $\nabla V(X_0) = 0$, then $F(X_0, 2\pi) = X_0$ and $\nabla V(X) = 0$ on the orbit of X_0.

In case of a 2π-periodic orbit, then $F(X_0, 2\pi) = X_0$ and $\nabla V(X_0)$ is orthogonal to *Range* $(I - \Phi(2\pi))$. In other words $\nabla V(X_0)$ belongs to *ker* $(I - \Phi(2\pi)^T)$ and generates it if it is non-zero and if *ker* $(I - \Phi(2\pi))$ is generated only by $\dot{X}_0(0)$. Furthermore in this case the algebraic multiplicity has to be more than 1. In fact, since $\dot{X}_0(0)$ is orthogonal to $\nabla V(X_0)$, then $\dot{X}_0(0)$ belongs to *Range* $(I - \Phi(2\pi))$ hence there is another vector in *ker* $(I - \Phi(2\pi))^2$ besides $\dot{X}_0(0)$.

Consider the problem of finding 2π-periodic solutions to the equation

$$\dot{X} = g(X) + \nu \nabla V(X) = g(X, \nu).$$

If $X(t)$ is such a solution, then $\dot{X} \cdot \nabla V(X) = \nu \, |\nabla V(X)|^2 = \frac{d}{dt} V(X(t))$. Integrating over a period one has $\nu |\nabla V(X)| = 0$, thus $\nu = 0$ if, on the orbit $\nabla V \neq 0$, or $\nabla V(X) = 0$ on the orbit and in both cases $X(t)$ is a 2π-periodic solution of the original problem. Furthermore, since $g_\nu(X_0, 0) = \nabla V(X_0)$, one has

$$\int_0^{2\pi} g_\nu \cdot Z(t) dt = \int_0^{2\pi} \nabla V(X_0(t)) \cdot \Phi(t)\Phi^{-1}(t)\Phi^{-1T}(t)\nabla V(X_0(0)) dt$$
$$= \int_0^{2\pi} \nabla V(X_0(0)) \cdot \Phi^{-1}(t)\Phi^{-1T}(t)\nabla V(X_0(0)) dt,$$

where one has used the relation $\nabla V(X(t)) \cdot \Phi(t)W = \nabla V(X(0)) \cdot W$.

Since $\Phi^{-1}\Phi^{-1T}$ is a positive definite matrix, the integrand is positive and the only condition for hyperbolicity in this case is that $dimker(I - \Phi(2\pi)) = 1$, or else whenever X is a 2π-periodic solution of $\frac{dX}{dt} - BX = 0$, then X is a multiple of $\dot{X}_0$.

Let ω be an open bounded subset of $\mathbb{R}^N$ such that any 2π-periodic orbit of $\dot{X} - g(X) = 0$ in $\bar{\omega}$ is in fact in ω. (No assumption on $g(X) \neq 0$ or $\nabla V(X) \neq 0$ is made at this stage).

If $X(t)$ is a 2π-periodic solution, with $X(t)$ in ω, then $|X(t)|$ is bounded in $C^1(S^1)$ and thus in $H^1(S^1)$ by some R. Let $\Omega = \{(X, \nu), X \in H^1(S^1), \| X \|_1 < R, |\nu| < \varepsilon, X(t) \in \omega$ for all $t\}$.

As in [I.M.V] it is easy to see that Ω is open and that any 2π-periodic solution of $\dot{X} - g(X) - \nu \nabla V(X) = 0$ (that is of $\dot{X} - g(X) = 0$) in $\bar{\Omega}$ is either in Ω or $\nabla V(X) = 0$ on it.

Proposition 5.9.

1) If $g(X)$ and $\nabla V(X)$ are non-zero in ω, then the Dancer-Toland degree is the norma-lized S^1-degree of [I.M.V].

2) The S^1-degree of $\dot{X} - g(X) - \nu \nabla V(X)$ is well defined with respect to Ω, provided $\nabla V \neq 0$ on 2π-periodic solutions (including stationary ones). In case of a hyperbolic orbit the S^1-index will have at most two non-zero components: $d_{H_0} = Sign\,(e_{k-1} \cdot \nabla V(X_0))(-1)^{k+\sigma_+}$ and $d_{H_1} = Sign\,(e_{k-1} \cdot \nabla V(X_0))(-1)^{k+\sigma_+}((-1)^{\sigma_-} - 1)/2$, if $|H_0| = 2|H_1|$.

5.8. Symmetry breaking for differential equations.

In this final section of this chapter we shall collect all the results on symmetry breaking obtained in an abstract form in Chapter IV as applications of perturbations near a periodic orbit. We urge the reader to have present the considerations given in Remark 5.3.

Assume that the equation $\frac{dX}{dt} = g(X, \nu)$ has an $\{e\}$-hyperbolic periodic solution (X_0, ν_0) of minimal period $2\pi/n_0$, as in Section 5.6. For example if $g(X, \nu) = g(X)/\nu$, the equation $X' = g(X)$ has X_0 as a periodic solution of minimal period $2\pi/\nu_0 n_0$.

From Proposition 5.8, $I_H = \eta(-1)^{n_H}$, for any divisor $|H|$ of n_0. Moreover, n_H has the parity of σ_+ (the number of Floquet multipliers of the first Poincaré return map which are real and larger than 1) if $n_0/|H|$ is odd. On the other hand, n_H has the parity of $\sigma_- + \sigma_+$ if $n_0/|H|$ is even (σ_- the number of Floquet multipliers which are real and less than -1).

Furthermore, from Proposition 5.2, the S^1-index at (X_0, ν_0) has at most two non-zero components: $d_0 = \eta(-1)^{\sigma_+}$ for $|H_0| = n_0$ and, if n_0 is even, $d_1 = \eta[(-1)^{\sigma_- + \sigma_+} - (-1)^{\sigma_+}]/2$ for $|H_1| = n_0/2$. If $|H|$ divides n_0, then the S^1-index in V^H, corresponding to the modes which are multiples of $|H|$, will be $d_0[F_{H_0}]_{S^1}$ if $n_0/|H|$ is odd and $d_0[F_{H_0}]_{S^1} + d_1[H_1]_{S^1}$ if $n_0/|H|$ is even.

Consider now the equation

$$(5.1) \qquad \frac{dX}{dt} = g(X, \nu) + \tau h(t, X, \nu),$$

for small τ and where h is $(2\pi/q)$-periodic in t.

For example one may consider the equation

$$(5.2) \qquad \frac{dX}{ds} = g(X) + \tau h(\omega s, X)$$

where $h(u, X)$ is 2π-periodic in u and $\frac{dX}{ds} = g(X)$ has $X_0(s)$ as a $2\pi/\omega_0$ periodic solution. It is clear that (5.2) will give the first equation with $g(X, \nu) = g(X)/\nu$ and $h(t, X, \nu) = h(qt, X)/\nu$ after the change of scale $t = \nu s$, with $\nu = \omega/q$. The unperturbed problem $\frac{dX}{dt} = g(X)/\nu$, has $X_0(t/\nu)$ as a $2\pi/n_0$ periodic solution provided $\nu = \nu_0 = \omega_0/n_0$, that is if $\omega_0/\omega = n_0/q$. This is an entrainment or phase locking problem, where one chooses n_0 and q relatively prime.

One has then the following result

Proposition 5.10. *Assume $X_0(t)$ is a solution of $\frac{dX}{dt} = g(X, \nu_0)$ such that the Kernel of $\frac{d}{dt} - Dg(X_0(t), \nu_0)$ is generated by $X_0'(t)$ and that the equation $\frac{dY}{dt} - Dg(X_0(t), \nu_0)Y = g_\nu(X_0(t), \nu_0)$ has no 2π-periodic solution. Let $|H|$ be the largest common divisor of n_0 and q, then (5.1) has a global continuum of $2\pi/|H|$- periodic solutions, starting from (X_0, ν_0) and parametrized by τ, if either $n_0/|H|$ is odd or, when $n_0/|H|$ is even, if σ_- is even.*

For (5.2) and a hyperbolic solution X_0, one has $2\pi q/\omega$-periodic solutions (for any small τ), starting from $\omega_0/\omega = n_0/q$ if either n_0 is odd or, when n_0 is even, if σ_- is even. The same results are valid in the non-hyperbolic case if Σd_j is odd.

Proof. It is enough to note that if $|H|$ divides q then $h(t, X, \nu)$ is $2\pi/|H|$ periodic. Thus, if $|H|$ divides also n_0, then one may consider the non-equivariant index of the equation (5.1) on V^H. This index will be $d_0 = \eta(-1)^{\sigma_+}$ if $n_0/|H|$ is odd and $d_0 + d_1 = \eta(-1)^{\sigma_+} + (1 + (-1)^{\sigma_-})/2$ if $n_0/|H|$ is even.

For (5.2), n_0 and q are relatively prime, then $|H| = 1$.

Q.E.D.

For planar systems, $\sigma_- = 0$ and one has a "curve" in each of the Arnold's tongues. For $n_0 = 1$, the solutions of (5.1) are called subharmonics of order q.

It would be quite interesting to have an example of a system of differential equations with, in a neighborhood of some set of periodic solutions, Σd_j even (and non zero) and such that this set is destroyed by a small non-autonomous perturbation. We shall give here an example for a pair of averaged Van der Pol's equations, that is for integro-differential equations. Look for 2π-periodic solutions to the equations:

$$x'' - (1 - \frac{1}{2\pi} \int_0^{2\pi} (x^2 + y^2) dt) x' + (1 + \nu)x + \tau(3\cos 2t\, y + \sin 2t\, y') = 0$$

$$y'' - (1 - \frac{1}{2\pi} \int_0^{2\pi} (x^2 + y^2) dt) y' + (1 + \nu)y - \tau(3\cos 2t\, x + \sin 2t\, x') = 0.$$

Here $q = 2$ and ν is close to 0. If $x(t) = \Sigma x_n e^{int}$, $y(t) = \Sigma y_n e^{int}$ and ρ^2 is the integral term, then for $n \geq 0$:

$$(-n^2 - in(1 - \rho^2) + 1 + \nu)x_n + \frac{\tau}{2}((n+1)y_{n-2} - (n-1)y_{n+2}) = 0$$

$$(-n^2 - in(1 - \rho^2) + 1 + \nu)y_n - \frac{\tau}{2}((n+1)x_{n-2} - (n-1)x_{n+2}) = 0.$$

For $\tau = 0$ and ν close to 0, the only nontrivial solutions are for $x_n = y_n = 0, n \neq 1, |x_1|^2 + |y_1|^2 = 1, \nu = 0$, corresponding to $x(t) = \alpha \cos(t + \varphi), y(t) = \beta \sin(t + \psi)$, with $\alpha^2 + \beta^2 = 2$. (The other solution is $x = y = 0$).

For $\tau \neq 0$, one has for $n = 1$:

$$(\nu - i(1 - \rho^2))x_1 + \tau \bar{y}_1 = 0$$

$$(\nu - i(1 - \rho^2))y_1 - \tau \bar{x}_1 = 0,$$

whose only solution is $x_1 = y_1 = 0$. The other equations form then a closed system with dominant diagonal terms for τ small and with a unique solution $x_n = y_n = 0$. Hence, for τ small and non-zero the only solution is $x = y = 0$.

It remains to compute the S^1-degree of the non-trivial solution for the autonomous problem. For $n \neq 1$, these coefficients can be deformed to 1 and the S^1-degree is that of

$((\nu - i(1 - \rho^2))x_1, (\nu - i(1 - \rho^2))y_1)$ or, after a rotation and the use of the suspension, the S^1-index of $(\nu - i(1 - |x_1|^2)^2)z_1$ near $|z_1| = 1$, which is 2.

For the bifurcation situation, if in (5.1), the map $g(x, \nu)$ and $h(t, x, \nu)$ depend on a parameter λ such that, for $\lambda \neq 0$, one has the $\{e\}$-hyperbolicity hypothesis, then one will have bifurcation for small τ (with this method) of a global continuum of $2\pi/|H|$-periodic solutions if $n_0/|H|$ is even and $\sigma_-(\lambda)$ changes parity. (See 4.5.2 and in particular Proposition 4.6). In contrast, with the result of Proposition 5.4, one will get such a bifurcation if $\sigma_+(\lambda)$ changes parity in case $n_0/|H|$ is odd or, in case $n_0/|H|$ is even, if $\sigma_-(\lambda)$ changes parity or, if $\sigma_-(\lambda)$ remains even and $\sigma_+(\lambda)$ changes parity.

For "degenerate" equations, then one will get bifurcation for small τ, if $\Sigma d_j(\lambda)$ changes parity. In the above coupled system, replace $1 - \rho^2$ by $\lambda(1 - \rho^2)$ or ν by $\nu\lambda$. In both cases the S^1-index will change from -2 to 2 as λ crosses 0, generating a bifurcation for any autonomous perturbation. However the only solution for $\tau \neq 0$ is $x = y \equiv 0$.

CHAPTER SIX

INDEX OF AN ISOLATED ORBIT OF STATIONARY SOLUTIONS AND APPLICATIONS.

We shall continue our computations of S^1-degrees in case of generic situations, but now when the set of solutions is a set of stationary solutions. This will lead us to Hopf bifurcations in several contexts.

Let $\Phi(x, z) : \mathbb{R}^M \times E \to \mathbb{R}^N \times E$ have 0 as a regular value and (x_0, z_0) in $(\mathbb{R}^M \times E)^{S^1}$ as a regular point. In the different cases we have studied so far, the first instance would be $k = l$, with x_0 an isolated zero. This is the standard generic situation for the Brouwer degree, hence we shall not treat it.

6.1. Computation of the S^1-Index.

For the case of $k = l + 1 - 2p$, if the map has 0 as a regular value, then there are no stationary zeros if $k \leq l - 1$. Thus, we shall restrict our attention to the case $k = l + 1$ and maps of the form $X - f(X, \nu)$ (hence $k_j = 1$, since for $k_j > 1$ it is not possible to linearize). Regular zeros form one-dimensional closed manifolds. Assume thus one has such a compact manifold, the simple loop $P \equiv (X_0(s), \nu(s))$, $s \in [0, 1]$, in $E^{S^1} \times \mathbb{R}$. This implies, from Theorem 4.1, that $dim\, E^{S^1} = l > 0$.

Since the linearization of f around P commutes with the action of the full group S^1, this linearization will not have cross - terms and will be of the diagonal form $D_{Z_j} f_j(X_0, \nu) = A_j(X_0, \nu)$ where the components of Z_j and of f_j have the unique mode m_j and A_j has a complex structure (see $[\mathbf{I_0}]$ and Section 4.5.3).

Thus, for $X = (X_0, Z_1, Z_2, \ldots)$ one has

$$X - f(X, \nu) = (X_0 - f_0(X_0, \nu) - g_0(X, \nu), \ldots, (I - A_j(X_0, \nu))Z_j + g_j(X, \nu), \ldots)$$

with $g_0(e^{i\varphi}X, \nu) = g_0(X, \nu) = 0(|Z|), g_j(e^{i\varphi}X, \nu) = e^{im_j\varphi}g_j(X, \nu) = 0(|X||Z|)$.

From the regularity one has that, on P, $X - D_{(X,\nu)}f$ has a one dimensional kernel generated by $(\dot{X}_0(s), \dot{\nu}(s))$. Furthermore this kernel is in the invariant component, that is $I - A_j|_P$ is invertible (and hence invertible in a neighborhood of P). Thus, in a neighborhood of P, $X - f(X, \nu)$ is S^1-homotopic to $(X_0 - f_0(X_0, \nu), \ldots, (I - A_j(X_0, \nu))Z_j, \ldots)$.

In the infinite dimensional case, the operator $A_j(X_0, \nu)$ is compact, thus one may use spectral projections on the eigenspaces with eigenvalues bigger than 1 and on an infinite-dimensional complement where $A_j(X_0, \nu)$ is deformable to 0. (Note that the same is true for $D_Z f(X_0, \nu)$, thus most of the $A'_j s$ are deformable to 0). Since P is compact one may do this reduction uniformly in the neighborhood of P. We shall assume, without loss of generality, that each Z_j is finite dimensional and that there is only a finite number of modes.

Furthermore the matrix $(I - A_j)\begin{pmatrix} a_j^{-1} & 0 \\ 0 & I \end{pmatrix}$, with $a_j(X_0, \nu) = det\ (I - A_j(X_0, \nu))$, represents, when restricted to P, a loop of complex matrices with determinant 1. Since this set is, by polar decomposition, homotopic to $SU(n)$ hence simply connected, this product is deformable to I on P and, by the covering homotopy extension property, on the neighborhood of P. Consequently f is deformable to

$$(X_0 - f_0(X_0, \nu), \ldots, \begin{pmatrix} a_j & 0 \\ 0 & I \end{pmatrix} Z_j, \ldots)$$

that is, via the Suspension Theorem 3.3,

$$\text{Index}_{S^1}(X - f(X, \nu); P) = \text{Index}_{S^1}(X_0 - f_0(X_0, \nu), \ldots, a_j z_j, \ldots; P)$$

where z_j is the first component of Z_j.

Given an orientation for P, the complex number $a_j(X_0(s), \nu(s))$ has a winding number ν_j, which will then characterize $I - A_j(X_0(s), \nu(s))$ as an element of Π_1 (G L $(\mathbb{C}^N)$). One has then the following result (compare with [**D.G.J.M**, Theorem 4.9]).

Proposition 6.1. *For the orbit* $(X_0(s), \nu(s))$, *the* S^1-*Index is*

$$d_0 = \quad \text{Index} \quad (X_0 - f_0(X_0, \nu); P) \ \text{in} \ \Pi_{l+1}(S^l) \ \text{and} \ d_j = -\eta \nu_j,$$

where η *is the sign of the determinant of the matrix* $\begin{pmatrix} I - f_{0X} & -f_{0\nu} \\ \dot{X}_0^T & \dot{\nu} \end{pmatrix}$

which is independent of s.

Proof. Note first that, by dividing by its norm, one may assume that $|a_j| = 1$. Let Ω_0 be a small neighborhood of P in $E^{S^1} \times \mathbb{R}$ and let $\Omega = \{(X_0, \nu, z_1, \ldots)/(X_0, \nu) \in \Omega_0, |z_j| < 4\varepsilon\}$. If $\psi(r) : \mathbb{R}^+ \to [0, 1]$ is a non-increasing function with value 1 for $0 \le r \le \varepsilon$ and value 0 for $r \ge 2\varepsilon$, define $\psi_j = \psi(|z_j|)$.

On $\partial\Omega$, the map $(X_0 - f_0(X_0, \nu), \ldots, a_j(X_0, \nu)z_j, \ldots)$ is S^1-homotopic to $\tilde{f}(X, \nu) = (X_0 - f_0(X_0, \nu), \ldots, (a_j(1 - \psi_j) + \psi_j)z_j, \ldots)$. This map has the form given in Proposition 4.3. Thus

$$d_j = \deg_{S^1}(\tilde{f}^{H_j}; \Omega_j),$$

with $\Omega_j = \Omega^{H_j} \setminus \cup_{0 \le k < j} \bar{B}_\varepsilon^k$, and d_j is obtained as an extension degree.

Now if $\tilde{f}(X, \nu) = 0$, then (X_0, ν) belongs to P, $z_j = 0$ or $a_j = -\psi_j/(1 - \psi_j)$ (real and negative) that is, since $|a_j| = 1$, $a_j = -1$, $\psi_j = 1/2$ and $|z_j| > \varepsilon$. It is clear that, over any finite number of modes, one may perturb $a_j(X_0, \nu)$ such that, over P, $a_j(X_0(s), \nu(s))$ attains the value -1 only a finite number of times, that it crosses the negative real axis with non-zero speed and that at these points on P, $a_k(X_0(s), \nu(s)) \ne -1$.

Thus if $\tilde{f}(X, \nu) = 0$, then either all z_j' s are zero or, if $|z| > \varepsilon$, only one of the z_j' s is non-zero. The corresponding zero $(X_0(s_0), \nu(s_0), z_j, \quad z_k = 0, k \ne j)$ has then the isotropy subgroup $\mathbb{Z}_{m_j}$. This implies that if H is an isotropy subgroup such that $|H| \ne m_k$ for all k, then any point in the corresponding Ω_H must have isotropy type H and consequently at least two non-zero components. Thus $\tilde{f}^H \ne 0$ on Ω_H and the component d_H is 0. While if $|H| = m_j$, then $\tilde{f}^H(X, \nu) = 0$ in Ω_j will imply that $z_j \ne 0, z_k = 0, k \ne j$: in fact if some other component is non zero (thus $z_j = 0$ from the above discussion), the isotropy type of the zero would be a multiple of m_j (recall that the modes are different).

Replacing $\psi_k = \psi(|z_k|)$ by $\psi(\tau|z_k|)$, for $k \ne j$, the above argument will give a S^1-homotopy of $\tilde{f}^H$ to $(X_0 - f_0(X_0, \nu), (a_j(1 - \psi_j) + \psi_j)z_j, z_k)$, on Ω_j. Then, from Theorem 3.3, $\deg_{S^1}(\tilde{f}^H; \Omega_j)$ is the S^1-degree of the map

$(X_0 - f_0(X_0, \nu), (a_j(1 - \psi_j) + \psi_j)z_j)$ on the set $\Omega_0 \times \{\varepsilon < |z_j| < 4\varepsilon\}$.

This S^1-degree is an extension degree which is computed by taking the restriction z_j real positive. It is the invariant degree of $(X_0 - f_0(X_0, \nu), a_j(1-\psi_j)+\psi_j)$ on $\Omega_0 \times \{\varepsilon < z_j < 4\varepsilon\}$, ($z_j$ can be deformed to 1).This degree is the sum of the indices at the points $(X_0(s_k), \nu(s_k))$ where $a_j(X_0(s_k), \nu(s_k)) = -1$, and $z_j = r_0$ is the unique point, between ε and 2ε, for which $\psi(r_0) = 1/2$. At such a point, one may perform the homotopy $((1-\psi_j)(tRea_j - (1-t))+ \psi_j, (1-\psi_j)Ima_j)$ and then deform to $(2\psi_j - 1, Ima_j)$. From the product theorem, the orientation of (X_0, ν, z_j) and the fact that ψ_j is decreasing, the index will be the index of $(X_0 - f_0(X_0, \nu), Ima_j)$ at $(X_0(s_k), \nu(s_k)) = P(s_k)$.

Choose a small neighborhood of $P(s_k)$ such that, one has for $s_k - \varepsilon \leq s < s_k$ the condition $(P(s) - P(s_k)) \cdot \dot{P}(s_k) < 0$ and, for $s_k < s \leq s_k + \varepsilon$, the condition $(P(s) - P(s_k)) \cdot \dot{P}(s_k) > 0$. (Parametrizing by arc - lenght, $\dot{P} \cdot \dot{P} = 1$). Consider the map

$$\Phi(X_0, \nu, s_k) = (X_0 - X_0(s_k)) \cdot \dot{X}_0(s_k) + (\nu - \nu(s_k))\dot{\nu}(s_k).$$

Then $(X_0 - f_0(X_0, \nu), \Phi(X_0, \nu, s_k))$ has an isolated zero at $P(s_k)$. Furthermore if Ima_j crosses the value 0 with positive speed, then a linear deformation from Ima_j to $\Phi(X_0, \nu, s_k)$ will give that the index at $P(s_k)$ is the index of $(X_0 - f_0(X_0, \nu), \Phi(X_0, \nu, s_k))$, while if Ima_j has negative speed then the index at $P(s_k)$ will be the index of

$(X_0 - f_0(X_0, \nu), -\Phi(X_0, \nu, s_k))$, that is the opposite of the index with $\Phi(X_0, \nu, s_k)$.

Freeing the parameter s_k and moving the neighborhood around $P(s_k)$ along with s_k, one sees that the index of the map $(X_0 - f_0(X_0, \nu), \Phi(X_0, \nu, s))$, at $P(s)$, is independent of s (it depends on the orientation of P). The linearization of this map at $P(s)$ is

$$\begin{pmatrix} I - f_{0X} & -f_{0\nu} \\ \dot{X}_0^T & \dot{\nu} \end{pmatrix}$$

Since the only element of the kernel of $(I - f_{0X}, -f_{0\nu})$ is $(\dot{X}_0, \dot{\nu})$, if the above matrix annihilates a vector (X, ζ), then $(X, \zeta) = k(\dot{X}_0, \dot{\nu})$ and $\dot{X}_0 \cdot X_0 + \dot{\nu}\zeta = k = 0$, i.e., the matrix is invertible. If η is the sign of its determinant (thus independent of s), the index at $P(s_k)$ is $\eta\, Sign\, d(Ima_j)/ds$. Since $d\,(Ima_j)/ds > 0$ contributes -1 to the winding number of a_j and $d(Ima_j)/ds < 0$ contributes $+1$, one gets the result.

Q.E.D.

Instead of perturbating $a_j(X_0, \nu)$ such that, over $P, a_j(X_0(s), \nu(s))$ attains the value -1 only a finite number of times with $a_k \neq -1$ at these points, one may use the following argument.

Assume that $|a_j| = 1$ and choose $k_j \in \mathbb{C}$, $|k_j| = 1$ with the following properties: consider the half lines $z_j(x) = k_j x/(1-x)$, for $0 \le x \le 1$, starting from 0 and going to ∞ in the direction k_j. Take the k_j's such that each line intersects the images $a_j(P)$ at only a finite number of points on P. By moving the lines one may achieve that the lines intersect $a_j(P)$ at different points of P. Then, at the beginning of the proof of Proposition 6.1, one may perform the homotopy

$$\left(X_0 - f_0(X_0, \nu), \ldots, ((1 - r\psi_j(|z_j|))a_j - rk_j\psi_j(|z_j|))z_j, \ldots\right)$$

If the corresponding map is 0, then (X_0, ν) belongs to P, $z_j = 0$ or $a_j = z_j(\tau\psi_j)$, with $\tau\psi_j = 1/2$ since $|a_j| = |k_j| = 1$. From the choice of the lines, this can happen at most for one j. The corresponding $|z_j|$ is less than 2ϵ and one is not on $\partial\Omega$. The rest of the argument is as before.

6.2. Application to bifurcation.

As an almost direct consequence of the previous computation and of Section 4.5.3, one may study the classical bifurcation of periodic solutions from a set of stationary solutions.

Suppose one has a family of S^1-maps $X - f(X, \nu, \lambda)$, such that $f(0, \nu, \lambda) = 0$. Assume that $I - f_X(0, \nu, \lambda)$ is singular at $(0,0)$ but invertible in a neighborhood of $(0,0)$. Let $A_j(\nu, \lambda) = f_{Z_j}(0, \nu, \lambda)$, as before, for $j \ge 0$. Hence, for at least one j, $I - A_j(0,0)$ is singular.

Zeros of the pair $(\varepsilon^2 - |X|^2, X - f(X, \nu, \lambda))$ will then be non trivial zeros of $X - f(X, \nu, \lambda)$ and, by considering the S^1-degree of this map, we shall be able to give results on local and global bifurcation as in Theorem 4.3 and Section 4.5.3. For a more complete study, see $[\mathbf{I_0}], [\mathbf{I_1}], [\mathbf{I.M.V_0}], [\mathbf{I.M.V}]$, etc. (See in particular Proposition 4.7 for "Necessary and Sufficient" conditions for S^1-bifurcation).

If $\Omega = \{(X, \nu, \lambda) : |X| < 2\varepsilon, \nu^2 + \lambda^2 < 4\rho^2\}$, one may choose $\varepsilon < \varepsilon(\rho)$ such that, if $X - f(X, \nu, \lambda) = 0$ for $|X| \le 2\varepsilon$ and $\nu^2 + \lambda^2 \ge \rho^2$, then $X = 0$. On $\partial\Omega$, one may then perform the S^1-homotopy $\tau(\varepsilon^2 - |X|^2) + (1 - \tau)(\nu^2 + \lambda^2 - \rho^2)$ and look at the S^1-map $(\nu^2 + \lambda^2 - \rho^2, X - f(X, \nu, \lambda))$ on Ω, with the isolated loop of zeros $\nu^2 + \lambda^2 = \rho^2, X = 0$. On this loop one may linearize $X - f(X, \nu, \lambda)$ to $((I - A_0(\nu, \lambda)X_0, \ldots, (I - A_j(\nu, \lambda)Z_j, \ldots)$.

Proposition 6.2. *Under the above hypothesis, then*

a). $d_0 = 0$ *if* $dim X_0 = 1$. *For higher dimension of* X_0, d_0 *is, in* $\mathbb{Z}_2$, *the suspension of the class of* $J(I - A_0)$, J *is the* J*-homomorphism. If* $d_0 \neq 0$ *then one has a global branch of stationary solutions. (If* $d_0 = 0$ *but* $J(I - A_0) \neq 0$, *i.e., only in the case* $dim X_0 = 2$, *one has a local branch).*

b). $d_j = (-1)^k$ *Sign det* $(I - A_0(\rho, 0))\nu_j$, *where* ν_j *is the winding number of* det $(I - A_j(\nu, \lambda))$. *If* $d_j \neq 0$ *then one has a global branch of bifurcating solutions in* E^{H_j} *(they may have a higher isotropy type).*

c). *If all the* d_j*'s are zero, then one may construct a nonlinearity without bifurcation.*

d). *For symmetry breaking situations, one has the results of Proposition 4.7.*

Proof. d_0 is the class of $(2t + 2\phi(X_0, \nu, \lambda) - 1, \nu^2 + \lambda^2 - \rho^2, (I - A_0)X_0)$, hence the suspension of the class of $(\nu^2 + \lambda^2 - \rho^2, (I - A_0)X_0)$ which is, up to an orientation, $J(I - A_0)$ (see [$\mathbf{I_1}$] for the study of this map and the application to bifurcation).

For d_j one has just to compute the value of η, the sign of the determinant of the

matrix $\begin{pmatrix} 0 & 2\nu & 2\lambda \\ I - A_0 & 0 & 0 \\ 0 & \dot\nu & \dot\lambda \end{pmatrix}$ with $\nu(s) = \rho\cos\ s$, $\lambda(s) = \rho\sin\ s$. This determinant is

$2\rho^2(-1)^l det(I - A_0)$. For the bifurcation result in this case see Theorem 4.3 and Proposition 4.7. Note that one may construct examples where the bifurcating branch is stationary, while if one knows that $I - A_0(0, 0)$ is invertible, then these solutions will have at least $\mathbb{Z}_{m_j}$ as isotropy subgroup. (c) and (d) are just a reminder of Proposition 4.7.

Q.E.D.

One may apply this result to the case of gradient maps and extend the results of [**Da**].

Remark 6.1. Note that one does not need to linearize $X_0 - f_0(X_0, \nu, \lambda)$ in order to compute d_j. In fact, from the proof of Proposition 6.1, it is enough to compute the index, at some point of P, of the map $(\nu^2 + \lambda^2 - \rho^2, X_0 - f_0(X_0, \nu, \lambda), \Phi(X_0, \nu, \lambda, s))$, assuming that P is an isolated orbit of stationary solutions.

For the bifurcation situation, $X_0 = 0, P = \{(\nu, \lambda)/\nu^2 + \lambda^2 = \rho^2\}$ and $\Phi(X_0, \nu, \lambda, s) = (\nu - \nu(s))\dot\nu(s) + (\lambda - \lambda(s))\dot\lambda(s)$. For a fixed s, $(\nu^2 + \lambda^2 - \rho^2, \Phi(X_0, \nu, \lambda, s))$ has an isolated zero at $(\nu(s), \lambda(s))$. Thus, near $(0, \nu(s), \lambda(s))$ one may deform $X_0 - f_0(X, \nu, \lambda)$ to $X_0 - f_0(X_0, \nu(s), \lambda(s))$ and use the product theorem. Hence

$$d_j = (-1)^k \, \text{Index} \, (X_0 - f(X_0, \nu(s), \lambda(s)); 0)\nu_j,$$

$$d_0 = [(\nu^2 + \lambda^2 - \rho^2, X_0 - f(X_0, \nu, \lambda)); \| \, X_0 \, \| \leq 2\epsilon, \nu^2 + \lambda^2 \leq 4\rho^2].$$

Remark 6.2. The following observation is valid for the application of the S^1-degree in any general context, where $k = l + 1$, but it might be better understood in the case of bifurcation. Assume that $\deg_{S^1}(f; \Omega) = \Sigma d_j [F_{H_j}]_{S^1}$. Then, from Remark 3.6, one has that $\deg_{S^1}(f^H; \Omega^H) = \Sigma_{H_j \geq H} d_j [F_{H_j}^H]_{S^1}$. Hence, if d_r and d_s are non-zero for two subgroups H_r and H_s, one has a solution for $f^H(x) = 0$ in Ω^H (and by equivariance for $f(x) = 0$) in Ω^{H_r} and Ω^{H_s}. In general, these two solutions may be the same, that is, if the common solution belongs to $\Omega^{H_r} \cap \Omega^{H_s}$ with an isotropy subgroup H which contains $H_r \cup H_s$.

Even if $d_H = 0$, this is not enough to guarantee that the two solutions are distinct. In fact, consider the following example:

$$E = \mathbb{R} \times \mathbb{C}^4, m_2 = 2, m_3 = 3, m_5 = m_6 = 6, k_j = 1 \text{ with the map}$$

$$(\lambda z_2, \lambda z_3, \lambda z_5 + z_2^3, \bar{\lambda} z_6 + z_3^2, |z_2|^2 + |z_3|^2 + |z_5|^6 + |z_6|^2 - \epsilon^2)$$

where $\lambda = \mu + i\nu$. That is, one looks for non trivial solutions of the first four equations, or else $\lambda = 0, z_2 = z_3 = 0, |z_5|^2 + |z_6|^2 = \epsilon^2$, i.e., a bifurcation in E^H, for $H = \mathbb{Z}_6$. d_6 is computed from the last three equations with $z_2 = z_3 = 0$ and is it clear that $d_6 = 0$: use the deformation $(\lambda z_5 + \tau z_6, \bar{\lambda} z_6 - \tau z_5)$. In order to compute d_3, put $z_2 = 0$, deform on $\partial U_\epsilon, z_3^2$ to 0, rotate the components z_5 and z_6 as above and get a suspension. One has to look at $(\lambda z_3, z_5, z_6)$ with degree 1. Thus, $d_2 = d_3 = 1$.

Similarly one may look at the map $(\lambda z_2, \lambda z_3, |\lambda|^2 x_0 + |z_2|^2 + |z_3|^2)$ for $|z_2|^2 + |z_3|^2 + |x_0|^2 - \epsilon^2 = 0$. Here $d_0 = 0$ (corresponding to S^1), $d_2 = d_3 = 1$ and the non-trivial solution are in E^{S^1}.

On the other hand, one has the following:

Proposition 6.3. *Assume that d_r and d_s are non-zero. Suppose that $f^H(x) \neq 0$ for x in Ω^H, for all H's which contain $H_r \cup H_s$. Then $f(x) = 0$ has two distinct solutions in Ω^{H_r} and Ω^{H_s}. In the case of the Hopf bifurcation these solutions will be different provided $I - A_j(0,0)$ is invertible for all j's common multiples of r and s and provided $X_0 - f_0(X_0, \nu, \lambda)$ has no other zero, near $\nu = \lambda = 0$, than $X_0 = 0$.*

The last part of the Proposition 6.3 will be handy when there is no bifurcation of stationary solutions and when the higher modes in the bifurcation equations have no strong resonances.

6.3. Hopf Bifurcation for Autonomous Differential Equations.

For the family

$$\frac{dY}{dt} = g(Y,\lambda) = L(\lambda)Y + h(Y,\lambda), \quad Y \in \ \mathbb{R}^m, \text{ with } g(0,\lambda) = 0,$$

one obtains the linearization, at the level of Fourier series, $(i\nu m_j I - L(\lambda))X_j$, X_j in $\mathbb{C}^m$, see [$\mathbf{I_0}$].

Assume that $L(0)$ has the eigenvalues $\pm i\nu_0 m_1,\dots,\pm i\nu_0 m_s$ and that $L(\lambda)$ has its corresponding eigenvalues off the imaginary axis for $\lambda \neq 0$, λ small. $L(\lambda)$ may also have 0 as an eigenvalue.

From the hypothesis it is clear that the matrices $i\nu m_j I - L(\lambda)$ are invertible for $(\nu - \nu_0, \lambda) \neq (0,0)$.

Since the invariant part $(m_0 = 0)$ is independent of ν, one cannot apply directly the previous argument, unless $L(0)$ is invertible.

However if one assumes that if $g(Y,\lambda) = 0$, for $|Y| < 2\varepsilon, |\lambda| < 2\rho$, then $Y = 0$, i.e., that there are no non-trivial stationary solutions, then one may consider the map $(\varepsilon^2 - |X|^2, X - f(X,\nu,\lambda))$ which is S^1-deformable on the set $\{|X_0| < 2\varepsilon, |Z| < 2\varepsilon, (\nu - \nu_0)^2 + \lambda^2 < 4\rho^2\}$ to $((\nu - \nu_0)^2 + \lambda^2 - \rho^2, X_0 - f_0(X_0,\lambda), (i\nu m_j I - L(\lambda))X_j)$. Here $|X|$ and $|Z|$ mean the H^1-norm for the Fourier series. See [$\mathbf{I.M.V_0}$, Proposition 4.7] for the details.

Proposition 6.4. *Under the above hypothesis, then one has:*

a). $d_0 = 0$

b). $d_j = $ Index $(g(X,0);0)\sigma_j$, where σ_j is the net crossing number of eigenvalues at $i\nu_0 m_j$, (see [$\mathbf{I_0}$]).

Proof. The invariant component of the Fourier series, evaluated at X_0 is just $X_0 - f_0(X_0,\lambda) = -g(X_0,\lambda)$. Then for d_0, one has to consider the map

$((\nu - \nu_0)^2 + \lambda^2 - \rho^2, -g(X_0, \lambda))$ which is deformable to $((\nu - \nu_0)^2 + \lambda^2 + \rho^2, -g(X_0, \lambda))$ on the boundary of the set $\{|X_0| \leq 2\varepsilon, (\nu - \nu_0)^2 + \lambda^2 \leq 4\rho^2\}$.

For d_j, the winding number of $det(I - A_j)$ gives minus the net crossing number of eigenvalues (see [I_0]).

The index, η, of the map: $(X_0, \nu, \lambda) \rightarrow ((\nu - \nu_0)^2 + \lambda^2 - \rho^2, -g(X_0, \lambda), \Phi(X_0, \nu, \lambda, s))$ where $\Phi(X_0, \nu, \lambda, s) = (\lambda - \rho sin \; s)\rho cos \; s + (\nu - \nu_0 - \rho cos \; s)(-\rho sin \; s)$ for $\nu - \nu_0 = \rho cos \; s$, $\lambda = \rho sin \; s$, is then Index $(g(X_0, 0); 0)$Index$((\nu - \nu_0)^2 + \lambda^2 - \rho^2, \Phi(\nu, \lambda, s))$, as shown in Remark 6.1. Since the last index is 1 (linearize), one gets the result.

$$Q.E.D.$$

Remark 6.3. Multiplicity results.

Besides the multiplicity results given in Proposition 6.3, one may, for the case of differential equations, apply Corollary 4.2 and its consequences. In fact, the fundamental matrix $\Phi(t)$ for $\nu_0 \frac{dX}{dt} = B(t)X$, where $B(t) = Dg(X(t))$ at a small solution $X(t)$ of $\nu_0 \frac{dX}{dt} = g(X)$, is close to $exp \; (Lt/\nu_0)$, where $L = Dg(0)$. Thus, the Floquet multipliers at $(X(t), \nu, \lambda)$ will be close to $exp \; (2\pi\mu(\lambda)/\nu)$, where $\mu(\lambda)$ are the eigenvalues of $L(\lambda)$. In particular if $\mu(\lambda) = \alpha \pm i\beta$, then the Floquet multiplier will be close to $exp \; (2\pi\alpha/\nu) \; exp \; (\pm 2\pi i\beta/\nu)$. Its real part will be negative only if β is an odd multiple of $\nu/2$. This will not be the case for $m_j\nu_0$. Then it is clear that one may replace ν_0 by $\nu_0/2$ or $\nu_0/2^k$ in such a way that, for the corresponding m_j's, no eigenvalue of $L(0)$ is an odd multiple of $\nu_0/2$. Then all Floquet multipliers for the corresponding solutions (iterated 2^k times) will have positive real part. Taking $k = 2$ and if $\alpha \neq 0$ for β odd multiple of $\nu_0/4$, then the eigenvalues with negative real part (for $\nu_0/2$) will come by pairs and will be away from 1. In all cases, by repeating the argument for q-roots of unity, one may choose ν_0 such that σ_- is even (or 0) for all solutions close to 0.

If there is no bifurcation of stationary solutions, and $Re\mu(\lambda) \neq 0$ if $\lambda \neq \lambda_0$ and $\mu(0) = im_j$, then hypotheses (a), (b), (d), of Corollary 4.2 are verified, since for $\nu = \nu_0 \pm \rho_1, in I - L(\lambda)/\nu$ is invertible. Then, from Corollary 4.2, there is either vertical bifurcation, i.e., 2π-periodic solutions of $\nu_0 dX/dt = g(X, 0)$ or, for each j, $d_j =$ Index $(g(X, 0))\sigma_j = \deg_{S^1}(\nu dX/dt - g(X, \rho_1); W) - \deg_{S^1}(\nu dX/dt - g(X, -\rho_1); W)$, where $W^{\pm} = \{X, \nu)/\epsilon_1 < \| X \|_1 \leq 2\epsilon, |\nu - \nu_0| < \rho_1\}$.

If it is known that all bifurcated solutions are hyperbolic (then as explained above $\sigma_- = 0$ and they are isolated), then each one has just one d_j different from 0 and in fact it is ±1. From the notes after Corollary 4.2., one has then at least $\sum |\sigma_j|$ different solutions. The non-hyperbolic case will be treated in Chapter VII. However we shall leave to the reader the implications of these computations for the corresponding results. One may also perturb g to be in a generic situation.

It is clear that one may extend the arguments of Proposition 6.4 to the case of the problem of finding 2π-periodic solutions bifurcating from a set of stationary solutions parametrized by two parameters (ν, λ). Assuming without loss of generality that the set of "trivial" solutions are $(X_0 = 0, \nu, \lambda)$, consider the equation:

$$\frac{dX}{dt} = L(\nu, \lambda)X + h(X, \nu, \lambda)$$

where $h(X, \nu, \lambda) = o(\| X \|)$ and $L(\nu, \lambda)$ is a two-parameter family of constant matrices with the following spectral properties

a). For $(\nu_0, \lambda_0), L(\nu_0, \lambda_0)$ has eigenvalues $\pm im_1 \pm im_2, \ldots, \pm im_s$ and possibly 0.

b). For (ν, λ), close but different from (ν_0, λ_0), $L(\nu, \lambda)$ has no eigenvalues on the imaginary axis.

c). $L(\nu, \lambda)X + h(X, \nu, \lambda) = 0$ for $\| X \| \leq 4\epsilon, (\nu - \nu_0)^2 + (\lambda - \lambda_0)^2 \leq 2\rho^2$, then $X = 0$.

Then, from Remark 6.1 and Proposition 6.2, one has:

Proposition 6.5. *Under the above hypothesis,*

$$d_0 = 0$$

$$d_j = \text{Index}\,(L(\nu_0, \lambda_0)X + h(X, \nu_0, \lambda_0); 0)\nu_j$$

where ν_j is the winding number of $\det (ijI - L(\nu, \lambda))$ and the index may also be computed at any other point (ν, λ) close to (ν_0, λ_0).

Assume that (c) is replaced by (c') :

$$L(\nu, \lambda)X + h(X, \nu, \lambda) = 0 \text{ for } \| X \| \leq 2\epsilon, \rho^2 \leq (\nu - \nu_0)^2$$
$$+ (\lambda - \lambda_0)^2 \leq 4\rho^2, \text{ then } X = 0.$$

Then the conclusions of Proposition 6.5 have to be replaced by

$$d_0 = [(\nu - \nu_0)^2 + (\lambda - \lambda_0)^2 - 2\rho^2, L(\nu, \lambda)X + h(X, \nu, \lambda)); \parallel X \parallel \leq 2\epsilon, (\nu - \nu_0)^2$$
$$+ (\lambda - \lambda_0)^2 \leq 4\rho^2]$$
$$= J(L(\nu, \lambda)) \ if \ L(\nu, \lambda) \ is \ invertible \ for \ (\nu - \nu_0)^2 + (\lambda - \lambda_0)^2 = \rho^2.$$
$$d_j = -\text{Index}\,(L(\nu_0, \lambda_0 + \rho)X + h(X, \nu_0, \lambda_0 + \rho); 0)\nu_j.$$

Note that (c') and (b) are enough to guarantee that the full equation has no solution (except $X = 0$) for $(\nu - \nu_0)^2 + (\lambda - \lambda_0)^2 \geq \rho^2$. In fact, from (b) the equations in X_j are uniquely solvable and, since the residual part $h(X_0, X_1, \ldots, \nu, \lambda) = 0(\parallel X \parallel |Z|)$, then, for $\parallel X \parallel = \epsilon$ small enough, $Z = 0$. One is left with the invariant part.

6.4. Hopf Bifurcation for Systems with First Integrals.

As an application of the last proposition we shall study the problem for

$$\frac{dX}{dt} = L(\lambda)X + h(X, \lambda) = g(X, \lambda),$$

for which one has a family of first integrals $V(X, \lambda)$. Thus, $\nabla V(X, \lambda)$ is orthogonal to $g(X, \lambda)$, for each fixed λ. As explained in Section 5.7, this problem is equivalent to finding 2π-periodic solutions to the equation:

$$\frac{dX}{dt} = L(\lambda)X + h(X, \lambda) + \nu \nabla V(X, \lambda),$$

where, if one has a solution with $\nabla V(X, \lambda) \neq 0$, then $\nu = 0$.

Assume there is a family $X(\lambda)$ such that $g(X(\lambda), \lambda) = 0$, $\nabla V(X(\lambda), \lambda) = 0$. Without loss of generality we may take $X(\lambda) \equiv 0$. Assume also that $L(0)$ has eigenvalues $\pm im_1, \ldots, \pm im_s$, with $0 < m_1 \leq \ldots \leq m_s$ are counted with multiplicity. $L(0)$ may also be singular. Let

$$\nabla V(X, \lambda) = H(\lambda)X + k(X, \lambda) \ \text{ with } \ k(X, \lambda) = o(|X|).$$

Thus $H(\lambda) = H(\lambda)^T$.

Let $\mu_j(\lambda) = \alpha_j(\lambda) + i\beta_j(\lambda)$ be the eigenvalues of $L(\lambda)$, for λ close to 0, such that $\alpha_j(0) = 0, \beta_j(0) = m_j$.

The following hypothesis will be used throughout this section:

$$a). \text{ If } \mu_j(\lambda) = im_j, \text{ for } \lambda \text{ close to } 0, \text{ then } \lambda = 0,$$

(H_j)

$$b). \, ker \, H(0) \cap ker \, (im_j I - L(0)) = \{0\}, \text{ for } j = 1, \dots, s.$$

One has then the following result

Proposition 6.6. *Hypothesis* (H_j) *is equivalent to* $im_j I - L(\lambda) - \nu H(\lambda)$ *is invertible for* $(\lambda, \nu) \neq (0,0), (\lambda, \nu)$ *close to* $(0,0)$.

Proof. It is clear that if $im_j I - L(\lambda) - \nu H(\lambda)$ is invertible, then for $\nu = 0$ we shall get (a), while, for $\lambda = 0$, we shall obtain (b). The converse implication will follow from the observation:

Lemma 6.1. $L(\lambda)^T H(\lambda) + H(\lambda)L(\lambda) = 0.$

Proof. Linearizing the relation $g^T(X, \lambda)\nabla V(X, \lambda) = 0$, near (X, λ), and with respect to X, one obtains the identity

$$Dg^T(X, \lambda)\nabla V(X, \lambda) + D^2 V(X, \lambda)^T g(X, \lambda) = 0.$$

A new linearization of this expression near $X = 0$, and taking into account that $g(0, \lambda) = \nabla V(0, \lambda) = 0$, will yield the relation given in Lemma 6.1.

$$\text{Q.E.D.}$$

End of the proof of Proposition 6.6: Consider the complex scalar product

$$\big((L(\lambda) + \nu H(\lambda) - \mu I)U, H(\lambda)U\big) = (H(\lambda)L(\lambda)U, U) + \nu \parallel H(\lambda)U \parallel^2 -\mu(U, H(\lambda)U),$$

where $\mu = \alpha + i\beta$. Since $(H(\lambda)L(\lambda)U, U) = -(L(\lambda)^T H(\lambda)U, U) = -(U, H(\lambda)L(\lambda)U)$, this expression is purely imaginary. On the other hand $H(\lambda)$ is symmetric and real, thus $(U, H(\lambda)U)$ is real.

Hence, if μ is an eigenvalue of $L(\lambda) + \nu H(\lambda)$, with corresponding eigenfunction U, one has

$$\nu \parallel H(\lambda)U \parallel^2 = \alpha(H(\lambda)U, U).$$

One also has: $\parallel (L(\lambda) + \nu H(\lambda) - \mu I)U \parallel^2 = \parallel (L(\lambda) - \mu I)U \parallel^2 + \nu^2 \parallel H(\lambda)U \parallel^2$
$+2\nu \, Re \, ((L(\lambda) - \mu I)U, H(\lambda)U) = \parallel (L(\lambda) - \mu I)U \parallel^2 + \nu^2 \parallel\parallel H(\lambda)U \parallel^2 - 2\nu\alpha(U, H(\lambda)U)$.

Hence, if im_j is an eigenvalue of $L(\lambda) + \nu H(\lambda)$, with corresponding eigenvector U, then $\alpha = 0, \nu \parallel H(\lambda)U \parallel = 0$, U is an eigenvector of $L(\lambda) - im_j I$. From (a), $\lambda = 0$ and, from (b), $H(0)U \neq 0$, thus $\nu = 0$.

Q.E.D.

We shall also need some information on the spectral behavior of $L(\lambda)$:

Lemma 6.2. *Assume (H_j) holds, then for small λ one has*

a). For any $k \geq 1$, $ker \, H(\lambda) \cap ker \, (L(\lambda) - \mu_j(\lambda)I)^k = \{0\}$.

b). If μ is an eigenvalue of $L(\lambda)$, so are $\pm\bar{\mu}$ and $-\mu$, with the same algebraic multiplicity.

c). If $i\beta$ is a simple eigenvalue of $L(\lambda)$ with corresponding eigenvector U, then

$(H(\lambda) \, U, U) \neq 0$.

d). If $i\beta$ is an eigenvalue of $L(\lambda)$, with generalized eigenspace $ker \, (L(\lambda) - i\beta I)^k$, then $H(\lambda)$ generates a non-degenerate quadratic form on this eignspace, with a well defined signature $\sigma_\beta(\lambda)$.

Proof. If *(a)* is false for $k = 1$, then there are sequences λ_n converging to 0, μ_n converging to im_j, U_n with norm 1 and a subsequence converging to U, such that $(L(\lambda_n) - \mu_n I)U_n = 0, H(\lambda_n)U_n = 0$. Taking limits, one will get a contradiction to (H_j).

(b). From Lemma 6.1, one has $H(\lambda)(L(\lambda) - \mu I)^k = (-1)^k (L(\lambda)^T + \mu I)^k H(\lambda)$. Thus, if μ and $\bar{\mu}$ are eigenvalues of $L(\lambda), -\mu$ and $-\bar{\mu}$ are eigenvalues of $L(\lambda)^T$, with eigenvectors $H(\lambda)U$ (non-zero by *(a)*), and hence of $L(\lambda)$.

Let U be in $ker \, (L(\lambda) - \mu I)^k$ and set $V = (L(\lambda) - \mu I)^\gamma U$, where $\gamma \leq k - 1$ is the largest integer n for which $(L(\lambda) - \mu I)^n U \neq 0$. Thus, $(L(\lambda) - \mu I)V = 0$. If $H(\lambda)U = 0$, then $H(\lambda)V = 0$, hence, from the case $k = 1$ in *(a)*, $V = 0$, which results in a contradiction. This proves *(a)* for the case $k > 1$.

Since $H(\lambda)$ is a one to one morphism from $ker\,(L(\lambda) - \mu I)^k$ into $ker\,(L(\lambda)^T + \mu I)^k$, the second space may be bigger than the first. By interchanging the roles of $L(\lambda)$ and $L(\lambda)^T$, these spaces have the same dimension and $H(\lambda)$ is an isomorphism between them. The algebraic multiplicity equality follows from standard arguments.

(c). If $i\beta$ is a simple eigenvalue, with corresponding eigenvector U, then, from *(a)*, $H(\lambda)U \neq 0$ generates $ker\,(L(\lambda)^T + i\beta I)$. If U is orthogonal to $H(\lambda)U$, then U belongs to *Range* $(L(\lambda) - i\beta I)$ and the algebraic multiplicity of $i\beta$ cannot be one.

(d). Recall that $H(\lambda)$ is an isomorphism from $ker\,(L(\lambda) - i\beta I)^k$ onto $ker\,(L(\lambda)^T + i\beta I)^k$. If P is the orthogonal projection of $\mathbb{C}^M$ onto $ker\,(L(\lambda) - i\beta I)^k$, then $PH(\lambda)P$ generates a symmetric bilinear form on $ker\,(L(\lambda) - i\beta I)^k$. Furthermore if $PH(\lambda)U = 0$, for some U in this space, then $(U, H(\lambda)V)$ for all V's in the space and, since $H(\lambda)$ is an isomorphism onto $ker\,(L(\lambda)^T + i\beta I)^k$, U belongs to *Range* $(L(\lambda) - i\beta I)^k$ which complements $ker\,(L(\lambda) - i\beta I)^k$, that is $U = 0$. Hence the quadratic form is non-degenerate.

$$\text{Q.E.D.}$$

For the proof of *(d)*, one could have used Theorem I.3.3 in [**G.L.R**]. This result will be used in our next Lemma.

Decompose orthogonally $\mathbb{R}^M$ as $V(\lambda) \oplus V(\lambda)^\perp$, where $V(\lambda) = ker\,H(\lambda)$. The relationship $L(\lambda)^T H(\lambda) + H(\lambda)L(\lambda) = 0$, will imply that

$$L(\lambda) = \begin{pmatrix} A(\lambda) & B(\lambda) \\ 0 & C(\lambda) \end{pmatrix}$$

$$\text{where} \qquad C(\lambda)^T H(\lambda) + H(\lambda)C(\lambda) = 0.$$

Since, on $V(\lambda)^\perp$, $H(\lambda)$ is invertible, one has $C(\lambda) = -H^{-1}(\lambda)C(\lambda)^T H(\lambda)$ which implies that *dim* $V(\lambda)^\perp$ is even. Furthermore any eigenvalue of $L(\lambda)$ is either an eigenvalue of $A(\lambda)$ or of $C(\lambda)$. If μ is an eigenvalue of $A(\lambda)$ and (H_j) holds, μ has to be different from $\mu_j(\lambda)$. In other words, $\mu_j(\lambda)$ are eigenvalues of $C(\lambda)$ but not of $A(\lambda)$.

For our next result, we shall need part of the proof of [**G.L.R**, Theorem I.3.3].

Lemma 6.3. *For each fixed λ, one may perturb $L(\lambda)$ to $\tilde{L}(\lambda)$ such that the relationship $L(\lambda)^T H(\lambda) + H(\lambda) L(\lambda) = 0$ is preserved during the perturbation. $\tilde{L}(\lambda)$ has all its purely imaginary eigenvalues simple and $H(\lambda)$ has the same signature on the union of the corresponding eigenspaces for $L(\lambda)$ and $\tilde{L}(\lambda)$.*

Proof. From our previous considerations, it is enough to look at $C(\lambda)$. In order to lighten the notation, we shall drop the λ dependence. Let $i\beta$ be an eigenvalue of $C(\lambda) \equiv C$ and let k be such that

$$ker\,(C - i\beta I)^k \oplus Range\,(C - i\beta I)^k = \mathbb{C}^{\,s}$$

s is even. Let P be the orthogonal projection onto $ker\,(C - i\beta I)^k$ and let F be $i^{k-1} PH(C - i\beta I)^{k-1} P$. It is easy to check that $F^* = \bar{F}^T = F$.

Furthermore if $(FX, Y) \equiv 0$ for all Y in $ker\,(C - i\beta I)^k$, then X is orthogonal to $ker\,(C^T + i\beta I)^{k-1}$ (recall that H is an isomorphism from $ker\,(C - i\beta I)^{k-1}$ onto $ker\,(C^T + i\beta I)^{k-1}$. Thus X belongs to $Range\,(C - i\beta I)^{k-1}$. Hence, if F were 0, the ascent of $C - i\beta I$ would be $2k - 1$ and not k. (For $k = 1, F = PHP$ was already studied above).

Thus, there is $\lambda_1 \neq 0$ and U_1 such that $FU_1 = \lambda_1 U_1$. One may normalize U_1 in such a way that $(FU_1, U_1) = \text{Sign } \lambda_1 \equiv \eta_1$.

Let $U_j = i^{j-1}(C - i\beta I)^{j-1} U_1, j = 1, \dots, k$. Then

$$(HU_j, U_l) = \begin{cases} (-1)^{k-1} \eta_1 & \text{if} \quad l + j = k + 1 \\ 0 & \text{if} \quad l + j > k + 1. \end{cases}$$

Define

$$V_1 = U_1 + a_2 U_2 + \dots + a_k V_k$$
$$V_j = i^{j-1}(C - i\beta I)^{j-1} V_1, j = 1, \dots, k,$$

where $a_2, \dots, a_k$ are obtained by setting $(HV_1, V_j) = 0$ for $j = 1, \dots, k - 1$.

Then $V_k = U_k$ and it is easy to verify that $\{U_j\}$ and $\{V_j\}$ are two sub-bases for $ker\,(C - i\beta I)^k$.

Since $U_j = i(C - i\beta I)U_{j-1}$ and $V_j = i(C - i\beta I)V_{j-1}$, on these vectors the matrix $C - i\beta I$ is in Jordan form, J, with 0 on the diagonal and $-i$ on the upper diagonal. On the $\{V_j\}$ basis H is $(-1)^{k-1}\eta_1 P_k$, where P_k is a matrix with 1 on the second diagonal and 0 otherwise, that is $(P_k)_{ij} = 1$ for $i + j = k + 1$.

By repeating this process for each eigenvalue of F and then by lowering $k - 1$ to $k - 2$ etc..., one obtains a basis for $ker\, (C - i\beta I)^k$, that is a change of variables T for which $C - i\beta I$ is in the above Jordan form, hence $C = T^{-1}JT$ and $H = T^* \Lambda T$, where Λ is composed of the $(-1)^{k_j-1}\eta_j P_{k_j}$, matrices ($k_j$ the dimension of the corresponding Jordan block).

Let $J(\tau)$ be the matrix with diagonal terms $i\beta + i\tau\epsilon_j, j = 1, \ldots, k$ for the above construction, where $\epsilon_1, \ldots \epsilon_k$ are small and different, and the upper diagonal are $-i(1 - \tau)$.

Let $C(\tau) = T^{-1}J(\tau)T$. Then $AC^*(\tau)H + HC(\tau) = T^*(J^*\Lambda + \Lambda J)T$.

It is then easy to verify that $J^*\Lambda + \Lambda J = 0$. Furthermore, since $(HU, U) = (\Lambda TU, TU)$, then H and Λ have the same signature. On the block given by P_k, $det\, (\Lambda - \lambda I) = (-1)^k(\lambda - \eta_1)^k$, thus, the signature of Λ on that block is $(Sign\, \eta_1)k$ and the signature of H will be $\sum (Sign\, \eta_j k_j)$ where k_j is the size of the Jordan block and η_j is obtained as above. Note finally that $C(1)$ has all its eigenvalues (perturbed from $i\beta$) simple.

$$\text{Q.E.D.}$$

By moving slightly λ, the simple eigenvalues of $\tilde{L}(\lambda)$ will remain on the imaginary axis (see Lemma 6.2 (b)) and on its eigenspace $(H(\lambda)U, U)$ will keep the same sign. This implies that the signature of $H(\lambda)$ remains constant.

Let $\sigma_j^+(\lambda)$ be the sum of the signatures of $H(\lambda)$ on the generalized eigenspace of $L(\lambda)$, with eigenvalues $i\beta, \beta$ close to m_j but β larger than m_j. $\sigma_j^-(\lambda)$ will correspond to those eigenvalues with β smaller than m_j. Then $\sigma_j^+(\lambda) + \sigma_j^-(\lambda) = \sigma_j(0)$ and $\sigma_j^+(\lambda)$ is constant for $\lambda > 0$, λ small, or $\lambda < 0$.

We shall also need the following hypothesis

(H_0'). *There are ϵ and ρ small positive constants, such that whenever $g(X, \lambda) = 0$ with $\| X \| \leq 2\epsilon$, then, if $\rho \leq |\lambda| \leq 2\rho, X = 0$ and if $|\lambda| < \rho$ then either $X = 0$ or $\nabla V(X, \lambda) \neq 0$.*

Hypothesis (H_0') implies that if $g(X, \lambda) + \nu\nabla V(X, \lambda) = 0$ then either $X = 0$ or $|\lambda| \leq \rho$ and $\nu = 0$. Clearly (H_0') will hold if (H_0) is valid. If $g(X, \lambda) = 0$ and $\nabla V(X, \lambda) = 0$, then

$L(\lambda)X + \nu H(\lambda)X + h(X,\lambda) + \nu k(X,\lambda) = 0$ for any ν. Take $\nu \neq 0$, so that $L(\lambda) + \nu H(\lambda)$ is invertible and one gets the bound for $\parallel X \parallel$.

Thus, one will be able to define the S^1-degree of the map $(\epsilon^2 - \parallel X \parallel^2, \dot{X} - g(X,\lambda) - \nu\nabla V(X,\lambda))$ provided (H_j) holds for $j \geq 1$. One is then in the position of using Proposition 6.5.

Proposition 6.7. *Let (H'_0) and (H_j) hold for all j's ≥ 1. Then*

$$d_j = -\operatorname{Index}(g(X,\rho);0)(\sigma_j^+(\rho) - \sigma_j^-(\rho)).$$

If (H_0) also holds, then $\operatorname{Index}(g(X,\rho);0) = Sign\ (det\ L(\rho))$.

Proof. From Proposition 6.5, $d_j = -\operatorname{Index}(g(X,\rho);0)\nu_j$, where ν_j is the winding number of $det\ (L(\lambda) + \nu H(\lambda) - im_j I)$. By changing basis to $V(\lambda) \oplus V(\lambda)^\perp$ (the determinant will not change), one may assume that this matrix has the form

$$\begin{pmatrix} A(\lambda) - im_j I & B(\lambda) \\ 0 & C(\lambda) + \nu H(\lambda) - im_j I \end{pmatrix}.$$

$A(\lambda) - im_j I$ is a path of invertible complex matrices, hence deformable to I. $B(\lambda)$ may be deformed to 0 without changing the spectrum. One may then consider $C(\lambda) + \nu H(\lambda) - im_j I$. Make the perturbation of $L(\pm\rho)$ to $\tilde{L}(\pm\rho)$, as in Lemma 6.3. Thus, one may assume that $L(\pm\rho)$ have all their eigenvalues on the imaginary axis simple and for them $(H(\pm\rho)U,U) \neq 0$.

Now, $det\ (L(\lambda) + \nu H(\lambda) - im_j I) = \Pi_1^M a_k(\nu,\lambda)$, where $a_k(\nu,\lambda)$ are the eigenvalues counted according to their multiplicity.

If $a_k(\nu,\lambda)$ corresponds to an eigenvalues μ of $L(0)$, with μ different from im_j, then $a_k(\nu,\lambda)$ will remain away from 0 and will not wind around 0: one may then deform it to $\mu - im_j$ and then to 1.

Since the winding number of the product is the sum of the winding numbers of each $a_k(\nu,\lambda)$ close to 0, it is enough to look at each of them. If $a_k(\nu,\lambda) = \alpha(\nu,\lambda) + i\beta(\nu,\lambda)$ corresponds to an eigenvector U, for which $HU \neq 0$ and $\nu \parallel HU \parallel^2 = \alpha(HU,U)$, one has the following: since $\alpha(\nu,\lambda)$ is continuous, it must be non-zero for $\nu \neq 0$ and it keeps the same sign for all ν's positive (or negative). If for $\nu = 0, \lambda = \pm\rho, \alpha(0, \pm\rho) \neq 0$, then $a_k(\nu, \lambda)$ will stay on the same half complex plane and its winding number will be 0.

Hence, $a_k(\nu, \lambda)$ crosses the imaginary axis at most twice, for $\nu = 0$ and $\lambda = \pm\rho$. If this is the case, then $a_k(0, \pm\rho)$ is a simple eigenvalue, for which $(H(\pm\rho)U, U) \neq 0$. Thus, as ν crosses 0 from negative values to positive values, $\alpha(\nu, \lambda)$ will crosses 0 in the same direction if $(HU, U) > 0$ and in the other if $(HU, U) < 0$. In this case (HU, U) has the same sign on the whole loop: the continuity of the eigenvector is used only near $\pm\rho$ where it is true.

Taking the orientation (ν, λ), the loop described by $a_k(\nu, \lambda)$ will give a winding number 0 if $\beta(0, \pm\rho)$ have the same sign, 1 if $(HU, U) > 0$ and $\beta(0, -\rho) < 0 < \beta(0, \rho)$ or if the three inequalities are reversed, -1 if only one of the two sets of inequalities is reversed.

Let $n_-(-\rho, > 0)$ be the number of imaginary eigenvalues of $L(-\rho)$ for which $(HU, U) > 0$ and which are below im_j. Moreover $n_+(-\rho, > 0)$, $n_-(-\rho, < 0), n_+(-\rho, < 0)$ are defined similarly. Let $a_+(> 0)$ be the number of eigenvalues of $L(\lambda)$, which have $(HU, U) > 0$, for $\lambda = \pm\rho$, and which cross im_j from below to above as λ goes from $-\rho$ to ρ. The numbers $a_-(> 0), a_+(< 0), a_-(< 0)$ are defined similarly.

Then

$$n_-(-\rho, > 0) = a_+(> 0) + \{\text{eigenvalues with } (HU, U) > 0, \text{ which stay below } im_j\}$$

$$n_+(-\rho, > 0) = a_-(> 0) + \{\text{eigenvalues with } (HU, U) > 0, \text{ which stay above } im_j\}$$

$$n_-(\rho, > 0) = a_-(> 0) + \{\text{eigenvalues with } (HU, U) > 0, \text{ which stay below } im_j\}$$

$$= n_-(-\rho; > 0) + a_-(> 0) - a_+(> 0)$$

$$n_+(\rho, > 0) = a_+(> 0) + \{\text{eigenvalues with } (HU, U) > 0, \text{ which stay above } im_j\} =$$

$$n_+(-\rho, > 0) + a_+(> 0) - a_-(> 0).$$

The winding number is then $a_+(> 0) - a_+(< 0) + a_-(< 0) - a_-(> 0)$.

But $a_+ - a_- = n_+(\rho) - n_+(-\rho) = n_-(-\rho) - n_-(\rho)$, for the case $(HU, U) > 0$ as well as for the case $(HU, U) < 0$.

Since $\sigma_j^+(\rho) = n_+(\rho, > 0) - n(\rho, < 0)$, then the winding number will be $\sigma_j^+(\rho) - \sigma_j^+(-\rho) = \sigma_j^-(-\rho) - \sigma_j^-(\rho)$.

$$\text{Q.E.D.}$$

Remark 6.4. If $L(\lambda) = (\lambda + \lambda_0)L$, such that L has eigenvalues $\pm im_j/\lambda_0$, with $\lambda_0 > 0$, then $\sigma_j^+(-\rho) = 0, \sigma_j^+(\rho) = \sigma_j$ the signature of H for im_j/λ_0.

Remark 6.5. (H_0') could be strenghtened to ask that if $g(X,\lambda) = 0$ or $\nabla V(X,\lambda) = 0$, for $|X| \le 2\epsilon$, then $X = 0$. This would imply that M is even or Index $(g(X,\rho);0) = 0$, in which case $d_j = 0$. In fact, Index $(\tau g \pm (1-\tau)\nabla V; 0)$ is well defined since g and ∇V are orthogonal and non-zero for $X \ne 0$. Then Index $(g;0) = $ Index $(\nabla V;0) = $ Index $(-\nabla V;0) = (-1)^M$ Index $(\nabla V;0)$.

In order to compute d_0, assume that (H_0) holds. Thus, d_0 is the class of $J(L(\lambda) + \nu H(\lambda))$ in $\Pi_{M+1}(S^M)$. Since the J-homomorphism is an isomorphism from $\Pi_1(GL(\mathbb{R}^M))$ onto $\Pi_{M+1}(S^M)$, it is enough to find the class of $L(\lambda) + \nu H(\lambda)$ in the first group (see [$\mathbf{I_0}$]).

Decompose $\mathbb{R}^M$ into *ker* $H(0) \oplus$ *Range* $H(0)$ and write

$$L(\lambda) + \nu H(\lambda) = \begin{pmatrix} L_1 + \nu H_1 & L_2 + \nu H_2 \\ L_3 + \nu H_3 & L_4 + \nu H_4 \end{pmatrix}$$

Note that *dim ker* $H(\lambda) \le$ *dim ker* $H(0)$. If we had equality, then we could use the previous decomposition, $L(\lambda) + \nu H(\lambda)$ would be similar to a matrix with $H_1 = H_2 = H_3 = 0, L_3 = 0 : L(\lambda) + \nu H(\lambda) = Q^{-1}(\lambda)A(\nu,\lambda)Q(\lambda)$ and the class of $L(\lambda) + \nu H(\lambda)$ would be that of $A(\nu,\lambda)$ (composition of matrices gives sums in $\Pi_1(GL(\mathbb{R}^M))$ and $[Q(\lambda)] = -[Q^{-1}(\lambda)]$).

Here $H_1(0) = H_2(0) = H_3(0) = L_3(0) = 0, H_2^T = H_3$ and $L_1 + \nu H_1$ is invertible for small (ν, λ). Multiply $(L_2 + \nu H_2)$ and $(L_3 + \nu H_3)$ by $\cos \tau$ and replace $L_4 + \nu H_4$ by $L_4 + \nu H_4 - \sin^2\tau \ (L_3 + \nu H_3)(L_1 + \nu H_1)^{-1}(L_2 + \nu H_2)$ in the above matrix. The invertibility of $L(\lambda) + \nu H(\lambda)$ implies that $L_4 + \nu H_4 - (L_3 + \nu H_3)(L_1 + \nu H_1)^{-1}(L_2 + \nu H_2)$ is invertible. Thus, if the deformed matrix is singular, this may happen only for $(\nu, \lambda) = (0,0)$. For $\tau = \pi/2$, the deformed matrix is in diagonal blocks:

$$\begin{pmatrix} L_1 + \nu H_1 & 0 \\ 0 & A(\nu,\lambda) \end{pmatrix} = \begin{pmatrix} L_1 + \nu H_1 & 0 \\ 0 & I \end{pmatrix} \begin{pmatrix} I & 0 \\ 0 & A(\nu,\lambda) \end{pmatrix}$$

The class of the matrix will be the sum of the classes. Since $L_1 + \nu H_1$ is invertible, it is deformable to $L_1(0)$ and its class is trivial. Hence one has to compute (via the suspension theorem) the class of $A(\nu, \lambda)$. Now

$$A(\nu, \lambda) = D(\lambda) + \nu E^T(\lambda) + \nu 0(\lambda)$$

where $D(\lambda) = L_4 - L_3 L_1^{-1} L_2$, $E^T(\lambda) = H_4 - H_3 L_1^{-1} L_2$ and $0(\lambda) \to 0$ as $\lambda \to 0$.

From the relation $HL + L^T H = 0$, one deduces that $ED + D^T E^T = 0$. Furthermore if for some U, $DU + \nu E^T U + \tau \nu 0(\lambda) U = 0$, then by multiplying by $E^T U$, one has

$$(EDU, U) + \nu \parallel E^T U \parallel^2 + \tau \nu (0(\lambda) U, E^T U) = 0.$$

The first term is 0 due to the antisymmetry of ED. Then, for λ small enough, one needs $\nu = 0$ (E^T is invertible) and U would be an eigenvector of $A(0, \lambda)$, hence $\lambda = 0$. Thus, the above deformation is valid. Finally since $E(\lambda)$ is invertible for λ small and it has a trivial class, one may replace $D + \nu E^T$ by $ED + \nu EE^T$. Let $B = ED$, then $B + \nu(\tau EE^T + (1 - \tau)I)$ is a valid deformation (recall that $(BU, U) = 0$, E is invertible and B is invertible for $\lambda \neq 0$).

Since $B + B^T = 0$, there is, for each λ, a transformation $T(\lambda)$ such that $B(\lambda) = T^T(\lambda)\Lambda(\lambda)T(\lambda)$, with $det\, T(\lambda) > 0$ and $\Lambda(\lambda)$ is in the canonical Jordan form with diagonal blocks of the form $\begin{pmatrix} 0 & \beta_j \\ -\beta_j & 0 \end{pmatrix}$. If one asks that all β_j's must be positive (for $\lambda \neq 0$) then one of them will possibly have an orientation factor $\eta(\lambda) = \pm 1$. (One may also have $\eta(\lambda) = 1$ but then $det\, T(\lambda)$ may be negative).

As in the proof of the previous proposition, one may use the intervals $(\rho, \rho + \epsilon)$ and $(-\rho - \epsilon, -\rho)$ to deform $B(\pm\rho)$ to $\Lambda(\pm\rho)$, by taking paths for $T(\pm\rho)$ to I and one may also deform $\Lambda(\pm\rho)$ so that they have the same β_j's. Consider then the deformation

$$\tau B(\lambda) + (1 - \tau)[\lambda^2(B(\rho) + B(-\rho))/\rho^2 + \lambda(B(\rho) - B(-\rho))/\rho]/2 + \nu I.$$

Since all these matrices (except I) are antisymmetric, if for U this deformation is 0, take the product by U and obtain $\nu \parallel U \parallel^2 = 0$, that is $\nu = 0$. For $\lambda = \pm\rho$, the deformation is $B(\pm\rho)$, which are invertible, hence $\nu = 0$.

For $\tau = 0$, one may deform linearly to

$$B(\rho) + B(-\rho) + \lambda(B(\rho) - B(-\rho))/\rho + \nu I.$$

After the deformation to the Jordan form, one has that if $\eta(\rho)\eta(-\rho) = 1$, then $B(\rho) = B(-\rho)$ and one gets $2B(\rho) + \nu I$ which is deformable to I and the class is 0. While, if $\eta(\rho)\eta(-\rho) = -1$, then there is one block where $B(\rho) = -B(-\rho)$ on it, while on the others one obtains $2B(\rho) + \nu I$. The result is then the suspension of

$$\pm \lambda \begin{pmatrix} 0 & \beta \\ -\beta & 0 \end{pmatrix} + \nu I$$

and in both cases one obtains the generator of $\Pi_1(GL(\mathbb{R}^M))$. We have proved:

Proposition 6.8. $d_0 \equiv 1$ *in* $\mathbb{Z}_2$ *if and only if* $\eta(\rho)\eta(-\rho) = -1$. *If dim ker* $H(\lambda)$ *is constant, then* $B(\lambda) = H(\lambda)C(\lambda)$.

Remark 6.6. Suppose that $B(\lambda) = f(\lambda)C + g(\lambda)D$, where $f(\lambda)$ is odd and $g(\lambda)$ is even; both are non-zero for $\lambda \neq 0$ small. Assume that $(CU, DU) = 0$ for all U's and that C is an isomorphism from *ker* D into itself. Then the class of $B(\lambda) + \nu I$ is

(dim ker $D)/2$ (mod 2). In fact if $B(\lambda)U = 0$ then either $\lambda = 0$, or $DU = CU = 0$ which implies $U = 0$. $B(\rho) + B(-\rho) = 2g(\rho)D$. and $B(\rho) - B(-\rho) = 2f(\rho)C$. Decomposing the space as *ker* $D \oplus$ *Range* D, one has $U = V \oplus W$ and $2g(\rho)DU + 2(\lambda f(\rho)/\rho)U + \nu U = 2g(\rho)D\,W + \nu W + 2f(\rho)(\lambda/\rho)CV + \nu V + 2f(\rho)(\lambda/\rho)CW$. It is easy to see that one may deform the last term to 0 (multiply by W, DW and CV). Hence we have to consider the class of the matrix:

$$\begin{pmatrix} 2f(\rho)(\lambda/\rho)C + \nu I & 0 \\ 0 & 2g(\rho)D + \nu I \end{pmatrix}$$

Since D is invertible on *Range* D, the second row is deformable to $(0\ \ I)$ and, after a suspension, one has the class of the matrix $2f(\rho)(\lambda/\rho)C + \nu I$ on *ker* D. The rest of the argument is just the application of the Jordan form for C on *ker* D, obtaining

(dim ker $D)$ $/2$ times the Hopf map.

Remark 6.7. One may also compute the S^1-index for a Hopf bifurcation from a stationary solution $X = 0$ at $\lambda = 0$, i.e. $g(0, 0) = 0$, with $\nabla V(0, 0) \neq 0$. Let $L(0) = Dg(0, 0)$, then $\nabla V(0, 0)$ belongs to *ker* $L^T(0)$. Assume *dim ker* $L(0) = 1$. Let $X = kW + V$ where W generates *ker* $L(0)$ and V is in *Range* $L^T(0)$. One sees easily that if

$$F(k, V, \nu, \lambda) = g(X, \lambda) + \nu \nabla V(X, \lambda), \text{ then}$$

$$D_{(V,\nu)}F(0, 0, 0, 0)(U, \mu) = L(0)U + \mu \nabla V(0, 0)$$

is an isomorphism. Thus one gets $V(k, \lambda)$ and $\nu(k, \lambda) = 0$ as a unique local solution. After a change of variables, one may choose a curve in this 2-dimensional surface and assume that $g(0, \lambda) = 0$. Then one will get the surface in the form $X(k, \lambda) = kW_\lambda + V(k, \lambda)$ where W_λ belongs to $ker\ D(0, \lambda)$ and $V(k, \lambda) = 0(k^2)$.

Let $A(k, \lambda) = D_X g(X(k, \lambda), \lambda)$. Then $W_\lambda + \frac{\partial V}{\partial k}$ generates $ker\ A^T(k, \lambda)$, which is locally one-dimensional, and $\nabla V(X(k, \lambda), \lambda)$ generates $ker\ A^T(k, \lambda)$.

Write $X_0 = X(k, \lambda) \oplus U_0(\lambda)$, with U_0 orthogonal to $\nabla V(X(k, \lambda), \lambda)$, then, at the Fourier series level, $X(t) = X_0 + Z(t) + \bar{Z}(t)$, one obtains:

$$-(A(k, \lambda)U_0(\lambda) + \nu \nabla V(X(k, \lambda), \lambda)) + 0(\nu \parallel U_0 \parallel + \parallel U_0 \parallel^2 + \parallel Z \parallel_1^2 + k^2) = 0$$

$$(inI - A(k, \lambda))X_n + 0(\nu|X_n| + |k||X_n| + \parallel Z \parallel_1 (\parallel Z \parallel_1 + \parallel U_0 \parallel + |k|)) = 0.$$

As before, the first equation is uniquely solvable for $(U_0(\lambda), \nu)$ in terms of (λ, k, Z), with $\parallel U_0(\lambda) \parallel = 0(|k|^2 + \parallel Z \parallel_1^2)$. Hence $\nu = 0$ on the set of solutions of the other equations.

Assume $in\ I - A(k, \lambda)$ is invertible for $(k, \lambda) \neq (0, 0)$.

Then, on the set $U_\epsilon \equiv \{\parallel U_0 \parallel < 2\epsilon, |\nu| < 2\epsilon, \parallel Z \parallel_1 < 2\epsilon, |k| < 2\epsilon, |\lambda| < 2\epsilon\}$, the above equations, complemented by $\parallel Z \parallel_1 - \epsilon$, have a well defined S^1-degree equal to:

$$\deg_{S^1}(\parallel Z \parallel_1 - \epsilon, -A(0, 0)U_0(0) - \nu \nabla V(0, 0), (inI - A(k, \lambda))X_n; U_\epsilon)$$
$$= Sign\ det(-A(0, 0)\ \nabla V(0, 0))\ \deg_{S^1}(\parallel Z \parallel_1 - \epsilon, (inI - A(k, \lambda))X_n; V_\epsilon)$$

where $V_\epsilon = \{\parallel Z \parallel_1 < 2\epsilon, |k| < 2\epsilon, |\lambda| < 2\epsilon\}$ and the determinants computed on $(Range\ A^T(0)) \times \mathbb{R}$. Then

$$d_0 = 0$$
$$d_n = \eta \sigma_n$$

where η is the sign of the above determinant and σ_n is the winding number of $det\ (inI - A(k, \lambda))$, when $k^2 + \lambda^2 = \epsilon^2$.

Remark 6.8. Global bifurcation. From the results in section 4.5, one will obtain a global bifurcation if one of the d_j is non-zero. This global branch will be of truly periodic solutions if there is no bifurcation of stationary solutions, for example if $A(0)$ is invertible ($d_0 = 0$ in this case) or if $X_0 = 0$ is an isolated solution of $g(X_0, \lambda) = 0$. Now if the branch meets a solution of $X' = g(X, \lambda)$ with $\nabla V(X, \lambda) = 0$ (stationary or not) then the branch will be automatically unbounded in its ν component.

This situation is more likely to happen for a stationary solution. In fact, since $Dg(X, \lambda)^T \nabla V(X, \lambda) + H(X, \lambda)g(X, \lambda) = 0$, where $H(X, \lambda)$ is the Hessian of V, then, if $\nabla V(X, \lambda) = 0$, one has $H(X, \lambda)g(X, \lambda) = 0$. In the natural situation where H is invertible, one will get that $g(X, \lambda) = 0$. Or, in other words, the set of points where $\nabla V(X, \lambda) = 0$, for fixed λ, is discrete. But, as we have seen in section 5.7, if $\nabla V(X, \lambda) = 0$ at some point of an orbit, it must be 0 on the whole orbit.

A way of avoiding the unboundedness of the branch through the ν component is to complement the equation $X' - g(X, \lambda) - \nu \nabla V(X, \lambda)$ by $dist\ (X, \lambda; S) - \epsilon$, where S is the set of solutions (periodic or stationary) of $X' = g(X, \lambda)$, with $\nabla V(X, \lambda) = 0$.

Note that if (H_j) and either (H_o) or (H'_o) hold then, in a neighborhood of $(0,0)$, S is just $(0, \lambda)$. In fact if, for some solution of $X' = g(X, \lambda)$, one has $\nabla V(X, \lambda) = 0$, then for the Fourier series one has:

$$(inI - L(\lambda))X_n - h_n(X_0, X_1, \ldots, \lambda) = 0$$

$$H(\lambda)X_n + k_n(X_0, X_1, \ldots, \lambda) = 0.$$

Thus, taking $\nu = \rho$

$$(inI - L(\lambda) - \nu H(\lambda))X_n = h_n(X, \lambda) + \nu k_n(X, \lambda).$$

By Proposition 6.6, the linear part is invertible and the non-linear part is of the order of $\|\ Z\ \|\|\ X\ \|$. Hence for $\|\ X\ \|$ small enough, one has $Z = 0$ and $g(X, \lambda) = 0$. The same argument for $n = 0$, in case (H_0) holds, or by hypothesis, if (H'_0) holds, will yield the result.

Hence $dist\ (X, \lambda; S) = \|\ X\ \|$ and one may use the computations of Propositions 6.7 and 6.8. Thus, if one of the d_j's is non zero, the global branch will be unbounded (in X or λ) or will meet a point of S (a stationary point if $S \cap \{\lambda = \lambda_0\}$ is discrete).

Remark 6.9. Index computation for a subset of S.

In case the global branch obtained before is bounded and this branch meets S in the set $S_0 \equiv \{(X, \lambda)/g(X, \lambda) = 0, \nabla V(X, \lambda) = 0\}$ we know that, if S_0 has a "nice" structure (similar to the one for $X = 0, \lambda$ close to 0), then the sum of the local S^1-indices is 0 (see Theorem 4.3). It may be convenient to relax the condition on the "nice" structure.

Let $S_1 = \{(X, \lambda) \in S_0/inI - Dg(X, \lambda)$ is **not** invertible for some $n \geq 0\}$,

i.e., S_1 is the subset of S_0 from which a Hopf or stationary bifurcation may occur.

Points in $S_0\backslash S_1$ are curves $X(\lambda)$, for which $g(X(\lambda), \lambda) = 0$.

(a) Let Σ_0 be a subset of S_1 which is bounded, path connected and such that Σ_0 has a 2ϵ-neighborhood N with the property that $S_1 \cap N = \Sigma_0$.

(b) Let $H(X, \lambda)$ be the Hessian of $V(X, \lambda)$ and $A(X, \lambda) = Dg(X, \lambda)$. Assume that $inI - A(X, \lambda) - \nu H(X, \lambda)$ is invertible on Σ_0, for all n, if $\nu \neq 0$.

Thus, for ϵ small enough and $\nu = \pm 2\rho$, $inI - A(X, \lambda) - \nu H(X, \lambda)$ is invertible on N and, for $|\nu| \leq 2\rho$, these matrices will be invertible on the intersection of the curves $X(\lambda)$ with ∂N, (there are only a finite number of curves in N).

Consider the map

$$G(X, \lambda, \nu) \equiv$$
$$(\| Z \|_1^2 + dist(X_0, \lambda; S_0)^2 - \epsilon^2, -g(X_0, \lambda) - \nu \nabla V(X_0, \lambda) - 0(\| Z \|_1),$$
$$(inI - A(X_0, \lambda) - \nu H(X_0, \lambda))X_n + 0(\| Z \|_1^2))$$

on the set $\tilde{N} \equiv N \times \{Z/ \| Z \|_1 < 2\epsilon\} \times \{\nu/|\nu| < 2\rho\}$.

If $dist(X_0, \lambda; \Sigma_0) = 2\epsilon$ and $dist(X_0, \lambda; S_0) < \epsilon$, then a zero of the map will be on $X_0 = X_0(\lambda), Z = 0, g(X, \lambda) = 0 = \nabla V(X_0, \lambda)$, and the first component will be $-\epsilon^2$. Hence the S^1-degree of the map with respect to the open set $\tilde{N}$ is well defined and equal to the S^1-degree of

$$(\| Z \|_1^2 + dist^2(X_0, \lambda; S_0) - \epsilon^2, -g(X_0, \lambda) - \nu \nabla V(X_0, \lambda), (inI - A - \nu H)X_n).$$

Now, on N, one may consider the map

$$F_\epsilon(X_0, \lambda) = (dist(X_0, \lambda; \Sigma_0) - \epsilon, -g(X_0, \lambda) - \nabla V(X_0, \lambda)).$$

On ∂N the first component is ϵ, hence the Brouwer degree of this map with respect to N is 0 and it is the sum of the indices at the zeros of the map. These zeros are located on the curves $X(\lambda)$ at points $(X_j(\lambda_j), \lambda_j)$ which are at a distance ϵ from Σ_0.

Since $Dg(X_j(\lambda_j), \lambda_j)$ is invertible, the only solutions of $g(X, \lambda) = 0$ near $(X_j(\lambda_j), \lambda_j)$ are $(X(\lambda), \lambda)$. If the distance of $(X(\lambda), \lambda)$ to Σ_0 decreases when λ passes λ_j, then the above map is locally deformable to $(\lambda - \lambda_j, -g(X_0, \lambda))$ with an index equal to Index $(g(X_0, \lambda_j))$. The orientation is given by (X_0, λ). While if the distance increases then the map is linearly deformable to $(\lambda_j - \lambda, -g(X_0, \lambda))$ with index equal to $-$Index $(g(X_0, \lambda_j))$.

Since the sum of the local indices is 0, one has an even number of paths.

We shall say that $(X_j(\lambda_j), \lambda_j)$ is an entrance point to Σ_0 if Index $(F_\epsilon(X_0, \lambda); X_j(\lambda_j), \lambda_j)$ *is positive and an exit point if this index is negative. We shall also talk about entrance and exit paths.*

Note that one may have an entrance point with Index $(g(X_0, \lambda_j))$ negative and such that the path $X_j(\lambda)$ leaves geometrically Σ_0. However since Σ_0 is path connected, one may join two of these points by a path in Σ_0. On it, by hypothesis (b), $A(X, \lambda) + \nu H(X, \lambda)$ is always invertible for $\nu \neq 0$. Hence the sign of the determinant of this matrix is constant on Σ_0.

This will also be true for the entry and exit points. But there $A(X(\lambda_j), \lambda_j)$ is invertible and if $(A + \nu H)U = 0$, then by multiplying by HU, one gets $(AU, HU) + \nu \parallel HU \parallel^2 = 0$, where the first term is purely imaginary. Thus $\nu HU = 0$ and $AU = 0$ which is not possible. Therefore the sign of $det\ (A + \nu H)$ is the sign of $det\ A(X(\lambda_j), \lambda_j)$ and this is the Index of g at λ_j. Thus, Index (g, λ_j) is constant and entrance points are all of the same type.

Although assumptions (a) and (b) are probably enough for our next result, we shall replace (b) by (b') :

$$\ker(i\beta I - A(X_0, \lambda)) \cap ker\ H(X_0, \lambda) = \{0\}$$
$$\text{for all } \beta \geq 0 \text{ and all } (X_0, \lambda) \text{ in } \Sigma_0 \cup \{X_j(\lambda), \lambda\}.$$

If $inU - AU - \nu HU = 0$, for (X_0, λ) in Σ_0, then multiplying by HU, one has, as above, that $\nu HU = 0$ and $inU - AU = 0$. Thus (b') implies (b).

For each pair of entrance -exit points, one may compute η_{EXIT}, η_{ENTRANCE}, as in Proposition 6.8 and define $d_{0j} \equiv 1$ in $\mathbb{Z}_2$ if and only if the product of these two orientations is -1.

From assumption (b') and the fact that $(X_j(\lambda_j), \lambda_j)$ belongs to $S_1 \backslash S_0$, hence $inI - A(X, \lambda)$ is invertible for all $n \geq 0$, one has that $H(X, \lambda)$ has signatures on the generalized eigenspaces of $i\beta I - A(X, \lambda)$.

For $(X_j(\lambda_j), \lambda_j)$ let $\sigma_n^+(\lambda_j)$ be the sum of the signatures for $\beta > n$.

Proposition 6.9. *If* $\deg_{S^1}(G(X, \lambda, \nu); \tilde{N}) = (d_0, d_1, d_2, \ldots)$ *then under the hypotheses* (a) *and* (b'), *one has:*

$$a) d_n = \sum_{j\,\text{EXIT}} \sigma_n^+(\lambda_j) \quad - \sum_{j\,\text{ENTRANCE}} \sigma_n^+(\lambda_j)$$
$$\beta) d_0 = \sum d_{0j}$$

where d_{0j} *are defined as above.*

Proof. Put the neighborhood N into a big box, $|\lambda| \leq R$, $\| X_0 \| \leq R$ and choose disjoint paths outside N, extending $(X_j(\lambda), \lambda)$ and going to some point $\| X_j \| \leq R$ and $\lambda = -R$ if $X_j(\lambda_j)$ is an entrance point or to $\lambda = R$ if $X_j(\lambda_j)$ is an exit point (recall that $M \geq 2$). Take a 2ϵ-neigborhood of each of the paths with a cross-section orthogonal to the path at $(X(\lambda_j), \lambda_j)$. On the tube, extend the map $(-g(X_0, \lambda) - \nu \nabla V(X_0, \lambda), (inI - A(X_0, \lambda) - \nu H(X_0, \lambda)) X_n)$ by defining it, on the cross-section at some point $(X(\lambda), \lambda)$, as the value of this map on the cross-section at $(X_j(\lambda_j), \lambda_j)$ (the path is assumed to be smooth and to be orthogonal, at $\lambda = \pm R$, to the set $\{\| X_0 \| \leq R\}$). Extend this map outside the tubes to the box in any continuous way in (X_0, λ, ν). Let $\hat{g}(X_0, \lambda, \nu, Z)$ be this new map and consider

$$G_R(X, \lambda, \nu) = (\| Z \|_1^2 + d^2(X_0, \lambda) - \epsilon^2, \hat{g}(X_0, \lambda, \nu, Z))$$

where $d(X_0, \lambda)$ is the distance from (X_0, λ) to $(S_0 \cap N) \cup \{\text{paths}\}$. The zeros of G_R are for $Z = 0, (X_0, \lambda)$ in N, hence $\deg_{S^1}(G_R; B_R) = \deg_{S^1}(G; \tilde{N})$, where $B_R = \{\| X_0 \| < R, \| Z \| < 2\epsilon, |\lambda| < R, |\nu| < 2\rho\}$.

Notice that if $\hat{g}_0(X_0, \lambda, \nu) = -\hat{g}(X_0, \lambda, \nu, 0)$, then the index of $\hat{g}_0(X, \pm R, 0)$ is well defined near the intersection of the external paths with the big box and it is always positive. In fact it is the index of g at $(X_j(\lambda_j), \lambda_j)$ if $X_j(\lambda_j)$ is an entrance point and one is approaching

Σ_0 when λ increases (in this case Index $(g; \lambda_j)$ is positive) and, if $X_j(\lambda_j)$ is an entrance point but one is going away from Σ_0 when λ increases, (Index $(g; \lambda_j)$ is negative) then the external path has to turn back in order to reach $\lambda = -R$ and one obtains the result from the orientation. A similar argument works at exit points.

Construct a path, from an entrance point $(X_j, -R)$ to an exit point (X_k, R), which is outside the box, but contained in $\| X_0 \| \le 2R, |\lambda| \le 2R$. The path joins $(X_j, -R)$ to $(\bar{X}_0, -R) = A, (X_k, R)$ to $(\bar{X}_0, R) = B$ and is a straight line $(\bar{X}_0, \lambda)$ between A and B. Choose different $\bar{X}_0$'s for different pairs of entrance-exit points. Remember that there are as many entrance points as exit points. Take 2ϵ-neigborhoods of the new paths.

On the piece between $(X_j, -R)$ and A, deform $\hat{g}_0(X, \lambda, \nu)$ to $J(A(X_j(\lambda_j), \lambda_j)X +$

$\nu H(X_j(\lambda_j), \lambda_j)X)$ where J is an orientation matrix with determinant$-$Index (g). Recall that index $(\hat{g}_0, -R) = 1, \hat{g}_0$ is linearizable and one wants that the index of $-J(A + \nu H)$ at A should be -1. Perform the same deformation between (X_k, R) and B.

Now, take a path $(X_0(s), \lambda(s)), s \in [\text{-}R, R]$, in Σ_0 between $(X_j(\lambda_j), \lambda_j)$ and $(X_k(\lambda_k), \lambda_k)$. On the tube between A and B define the map $\tilde{g}(X + \bar{X}_0, \lambda, \nu, Z)$ as

$$(-J(A(X_0(s), \lambda(s))X + \nu H(X_0(s), \lambda(s))X),$$
$$(inI - A(X_0(s), X(s)) - \nu H(X_0(s), \lambda_s))X_n)$$

for $\lambda = s$.

Extend the maps obtained for the different tubes in any continuous way, to the ball $\{\| X_0 \| < 2R, |\lambda| < 2R, |\nu| < 2\rho, \| Z \|_1 < 2\epsilon\} = B_{2R}$.

What we are doing is to turn the map, on the path in Σ_0, inside-out. Now if $\tilde{d}(X_0, \lambda)$ is the distance from (X_0, λ) to $(S_0 \cap N) \cup \{\text{all internal and external paths}\}$, then $\deg_{S^1}(\| Z \|_1^2 + \tilde{d}(X_0, \lambda)^2 - \epsilon^2, \tilde{g}; B_{2R})$ is well defined. The zeros of the map will be either in N or in the tubes joining the points A and B. On the boundary of B_{2R} one may perform the deformation $inI - \tau(A + \nu H)$, since, if for some $n > 0, inX_n = \tau(A + \nu H)X_n$ then, multiplying by HX_n, one has $\nu\tau \| HX_n \|^2 = 0, in(X_n, HX_n) - \tau(AX_n, HX_n) = 0$ (by separating real and imaginary part). But then $\nu\tau HX_n = 0$ and $AX_n = (in/\tau)X_n$. On the boundary of B_{2R}, the first component is positive, except if $|\nu| = 2\rho$. Then, from hypothesis (b'), one has $X_n = 0, Z = 0$ and no solutions of the pair $(\tilde{d}(X_0, \lambda)^2 - \epsilon^2, \tilde{g}_0(X_0, \lambda, \nu))$.

Thus, $\deg_{S^1}(\| Z \|_1^2 + \tilde{d}(X,\lambda)^2 - \epsilon^2, \tilde{g}; B_{2R})$ is a suspension and is reduced to the class of its invariant part $[\tilde{d}(X_0,\lambda)^2 - \epsilon^2, \tilde{g}_0] = \tilde{d}_0$.

This implies that $\tilde{d}_n = 0$ for this map, or else that for d_n, defined from G and N, one has

$$d_n = -\sum d_{nT},$$

where the sum is over all the external tubes from the points A to the points B and the corresponding map is

$$(-J(A(X_0(s),\lambda(s)) + \nu H(X_0(s),\lambda(s)))X, (inI - A(X_0(s),\lambda(s)) - \nu H(X_0(s),\lambda(s)))X_n)$$

complemented by $\| Z \|_1^2 + \| X \|^2 - \epsilon^2$.

From Proposition 6.7 one obtains the result, since Index $(JA) = -1$ and that we have changed the sign of the complementary function.

In order to compute $\tilde{d}_0$, let Ω be the union of N with the tubular neighborhoods of all the paths: on $\partial\Omega$, $\tilde{d}(X_0,\lambda) = 2\epsilon$. Hence $\tilde{d}_0 = \deg(\tilde{d}(X_0,\lambda)^2 - \epsilon^2, \tilde{g}_0; \Omega \times \{\nu/|\nu| < 2\rho\})$ belongs to $\mathbb{Z}_2$. Recall that, from the construction, $\tilde{g}_0(X_0,\lambda,\nu) = 0$ implies that $\tilde{d}(X_0,\lambda) = 0$ or $\nu = 0$, both if (X_0,λ) is not in N or between A and B.

One may perform the deformation $\tau(\tilde{d}(X_0,\lambda)^2 - \epsilon^2) + (1-\tau)(\rho^2 - \nu^2)$. In fact, on $\partial\Omega$, if $\tilde{g}_0 = 0$ then $g(X_0,\lambda) = 0$ and $\nabla V(X_0,\lambda) \neq 0$ hence $\nu = 0$, or $A(X_0,\lambda)X = 0, H(X_0,\lambda)X \neq 0$ (hypothesis (b')) and again $\nu = 0$; that is the first component is positive. If $|\nu| = 2\rho$ and $\tilde{g}_0 = 0$ then $g(X_0,\lambda) = 0$, $\nabla V(X_0,\lambda) = 0$ and one is in Σ_0 or in one of the paths, with $\tilde{d}(X_0,\lambda) = 0$: the first component is negative. Thus

$$\tilde{d}_0 = [\rho - \nu, -g - \nu\nabla V]_{\Omega \times \{0<\nu<2\rho\}} + [\rho + \nu, -g - \nu\nabla V]_{\Omega \times \{-2\rho<\nu<0\}}$$
$$= \Sigma[-g + \rho\nabla V]_\Omega - \Sigma[-g - \rho\nabla V]_\Omega.$$

Now $\deg(-g + \rho\nabla V; \Omega) = \deg(-g + \rho\nabla V; N) + \sum_{AB} \deg(-g + \rho\nabla V; N_{AB})$ where N_{AB} are the tubes between the points A and B. In fact $-g + \rho\nabla V$ is non zero outside these sets. Since on the union of the N_{AB} one has streched out the set N with a change of orientation for the maps, it is not difficult to see that $\deg(-g + \rho\nabla V; \Omega) = 0$ in $\mathbb{Z}_2$.

Thus $\tilde{d}_0 = 0$ and one obtains the result.

Q.E.D.

Note that it is easy to modify the arguments given in the proof if, instead of having (b) valid also for $n = 0$, one has the equivalent of (H'_0) on Σ_0. This is the reason why we have defined the notion of entrance and exit points in a more general manner than strictly needed.

Note also that, if there is a bounded open set Ω_0 in $\mathbb{R}^M \times \mathbb{R}$ such that $\mathcal{S}_0 \cap \partial\Omega_0 = \emptyset$ (hence g or ∇V are not both zero on $\partial\Omega_0$), then one may define the S^1-degree of the map

$$G(X, \lambda, \nu) = (\|\, Z\, \|_1^2 + dist\,(X_0, \lambda; \mathcal{S}_0)^2 - \epsilon^2, dX/dt - g(X, \lambda) - \nu\nabla V(X, \lambda))$$

with respect to $\Omega_0 \times \{\|\, Z\, \|_1 < 2\epsilon\} \times \{|\nu| < 2\rho\} = \tilde{\Omega}$, provided ϵ is small enough and if $X(t)$ is a solution (close to $\mathcal{S}_0$) of the differential equation for $\nu \neq 0$ then $X(t) = X_0 \in \mathcal{S}_0$. This will be the case if hypothesis (b') is valid on $\mathcal{S}_0$. Also, in this case the deformation $inI - \tau(A(X_0, \lambda) + \nu H(X_0, \lambda))$ is possible on $\partial\tilde{\Omega}$, by repeating the argument of Proposition 6.9. Thus $\deg_{S^1}(G; \tilde{\Omega}) = \tilde{d}_0$. Furthermore it is also easy to show that $\tilde{d}_0 = \deg(-g + \rho\nabla V; \Omega_0) - \deg(-g - \rho\nabla V; \Omega_0)$. This difference will be 0 if $g \neq 0$ on $\partial\Omega_0$ or $\nabla V \neq 0$ on $\partial\Omega_0$. In any case $\tilde{d}_n = 0$ and Ω_0 cannot be connected to just one Hopf bifurcation point, with some $d_n \neq 0$.

Remark 6.10. Hamiltonian Systems.

Consider the system

$$\frac{dX}{dt} = J\nabla V(X, \lambda) \quad X \in \mathbb{R}^{2M}, J = \begin{pmatrix} 0 & -I \\ I & 0 \end{pmatrix}$$

or the case $\nabla V(X)/\lambda$ (λ will be the frequency), $\nabla V(0, \lambda) = 0$. Let $H(\lambda)$ be the Hessian of V at $(0, \lambda)$. Assume that $JH(\lambda_0)$ has eigenvalues $\pm im_1, \ldots, \pm im_s$.

Hypothesis (H_n) is met provided $JH(\lambda) - im_j I$ is invertible for $\lambda \neq \lambda_0$ (λ close to λ_0), hence always for the case $\nabla V/\lambda$.

Hypothesis (H'_0) is satisfied provided $X = 0$ is an isolated zero of $\nabla V(X, \lambda)$. This is the case if (H_0) holds, that is if $H(\lambda)$ is invertible for λ close to λ_0. Note that if $JHX =$

$(\alpha + i\beta)X$ then $(JHX, JX) = (HX, X) = \lambda(X, JX)$. Since (HX, X) is real and (X, JX) is pure imaginary, then either $\alpha = 0$ or $(HX, X) = 0 = (JX, X)$. Furthermore $JH - i\beta I = J(H + i\beta J)$ and $H + i\beta J$ is self-adjoint. If J conmutes with H (that is $V(P, Q) = V_0(P) + V_0(Q) + 0(|P|^2 + |Q|^2)$) then, it is easy to compute the spectrum of JH from that of H_0. If one is in the situation of [**F.R**, Theorem 8.4] then one has that $\mathbb{R}^M = E_1 \oplus E_2$ where E_j are invariant under the flow of JH, that all solutions of $\frac{dX}{dt} = JHX/\lambda_0$ with initial data in E_1, are 2π-periodic, while no solutions of this variational equation with initial data in $E_2 \backslash \{0\}$ are 2π-periodic and that there are no stationary solutions in $E_1 \backslash \{0\}$. It is easy to see that these hypotheses imply that JH splits into two matrices, $(JH)_j$, that $in\lambda_0 I - (JH)_2$ is invertible for all n, while $(JH)_1$ is invertible and has only pure imaginary eigenvalues $\pm im_1\lambda_0, \ldots, \pm im_s\lambda_0$ with algebraic multiplicity equal to the geometric multiplicity. In this case one may also verify that if σ_j is the signature of H on $ker(im_j I - JH)$ then $\sum \sigma_j = \nu/2$ where ν is the signature of H on E_1.

In general if $X = 0$ is an isolated zero of $\nabla V(X, \lambda)$ (in particular if $H(\lambda_0)$ is invertible) and if $im_j I - H(\lambda)$ is invertible for λ close to λ_0, but different from λ_0, (always true if $\nabla V(X, \lambda) = \nabla V(X)/\lambda$), then $d_j = \text{Index}(\nabla V(X, \lambda_0))(\sigma_j^+(\lambda_0 - \rho) - \sigma_j^+(\lambda_0 + \rho))$ (see Proposition 6.7). If $H(\lambda_0)$ is invertible, then $d_0 = 0$. If $\nabla V(X, \lambda) = \nabla V(X)/\lambda$, then $d_j = \text{Index}(\nabla V(X, \lambda_0))\sigma_j = Sign \det H(\lambda_0)\sigma_j$ if $H(\lambda_0)$ is invertible.

In this case, $\nabla V(X, \lambda) \neq 0$ on truly periodic small solutions. Hence, if isolated, they have a well defined S^1-index.

With the same arguments as in Remark 6.3, if there is no vertical bifurcation (for λ_0) and if these periodic solutions are hyperbolic in the sense of Proposition 5.9, then by a judicious choice of period, σ_- will be even and d_{H_0} will be the only non-zero component with $d_{H_0} = \pm 1$. From Corollary 4.2 and Remark 6.3, one will get at least $\sum |\sigma_j|$ different solutions (hence at least $|\nu/2|$). This will be the case for a generic perturbation of V or one may apply the computations of next Chapter. We leave, to the reader, the task of pursuing these multiplicity results and of comparing them to those of [**F. R**].

6.5. Hopf Bifurcation and Symmetry Breaking.

Again, as a last application of this chapter we shall consider the following problem.

Assume that the differential equation

$$\frac{dY}{ds} = L(\lambda)Y + h(Y,\lambda) = g(Y,\lambda)$$

with $g(0,\lambda) = 0$, satisfies the hypothesis of Proposition 6.4.

Perturb the equation by the function $\tau f(\nu s, Y, \lambda)$, where f is 2π-periodic in the first variable. By the change of scale $\nu s = t$, one obtains the equation

$$\nu \frac{dY}{dt} = g(Y,\lambda) + \tau f(t,Y,\lambda),$$

which is a particular case of the differential equation

$$\frac{dX}{dt} = L(\lambda,\nu)X + h(X,\lambda,\nu) + \tau f(t,X,\lambda,\nu).$$

Assume that

(a) $L(\lambda_0,\nu_0)$ has $\pm im_1,\ldots,\pm im_s$ as eigenvalues,

(b) $L(\lambda,\nu)$ has no eigenvalues on the imaginary axis, for (ν,λ) different from (λ_0,ν_0),

(c) $L(\lambda_0,\nu_0)$ is invertible,

(d) $f(t,0,\lambda,\nu) = 0$.

Under these assumptions the system

$$(\epsilon^2 - \parallel X \parallel^2, (inI - L(\lambda,\nu))X_n - h_n - \tau f_n)$$

has no zeros on the boundary of the set $\{\parallel X \parallel < 2\epsilon, (\nu - \nu_0)^2 + (\lambda - \lambda_0)^2 < \rho^2\}$, for appropriate ϵ and ρ, and for τ small enough.

Note that under more stringent hypothesis on f one could allow $L(\lambda_0,\nu_0)$ to be singular. The problem is that, since f is not equivariant, for $(\nu - \nu_0)^2 + (\lambda - \lambda_0)^2 = \rho^2$ one may solve X_n in function of X_0 and τ, but then $X_n(X_0,\tau)$ is not necessarily 0 as in the equivariant case. Thus one should assume that, for $(\nu - \nu_0)^2 + (\lambda - \lambda_0)^2 = \rho^2$, the original equation has no small solution but 0. At this stage this sort of hypotheses seems very strong, especially if one wants to treat the differential equation with $\frac{dY}{ds}$!

isotropy subgroup H_j, with $H < H_j$, will involve a subset of the modes m_j, and thus $|H_j|$ will be a multiple of $|H'|$). This is consistent with the Poincaré–section definition.

Decompose then $|H_0|$ in prime factors, $|H_0| = p_1^{\alpha_1} \dots p_k^{\alpha_k} \equiv P^\alpha$. Then $|H| = P^{\alpha-\beta}$ with $0 \le \beta \le \alpha$ (meaning $0 \le \beta_j \le \alpha_j$ for each j).

Denote I_H by I_β^α and d_H by $d_{\alpha-\beta}^\alpha$. Let $\Gamma_\alpha = \{(\delta_1, \dots, \delta_k), \delta_j = 0 \text{ or } 1\}$.

Our first task is to invert the matrix A of 5.2.

Lemma 7.1.

$$I_\beta^\alpha = \sum_{0 \le \gamma \le \beta} P^{\beta-\gamma} d_{\alpha-\beta+\gamma}^\alpha$$

$$P^\beta d_{\alpha-\beta}^\alpha = \sum_{\delta \in \Gamma_\alpha} (-1)^{|\delta|} I_{\beta-\delta}^\alpha$$

where $|\delta|$ is the number of δ_j's equal to 1 and $I_{\beta-\delta}^\alpha$ is taken to be 0 if $\beta_j < \delta_j$ for some j.

Proof. Since the relation between the I's and the d's is invertible, it is enough to verify that the substitution of the second identity into the first will yield $I_\beta^\alpha = I_\beta^\alpha$.

In fact note that $\Gamma_\alpha = (0, \Gamma_{\tilde\alpha}) \cup (1, \Gamma_{\tilde\alpha})$, where $\Gamma_{\tilde\alpha} = \{(\delta_2, \dots, \delta_k), \delta_j = 0 \text{ or } 1\}$. (thus Γ_α has 2^k elements). Then

$$P^\beta d_{\alpha-\beta}^\alpha = \sum_{\delta \in \Gamma_{\tilde\alpha}} (-1)^{|\delta|} (I_{\beta-\delta}^\alpha - I_{\beta-\delta-(1,0)}^\alpha)$$

$$\text{and} \quad \sum_{0 \le \gamma \le \beta} P^{\beta-\gamma} d_{\alpha-\beta+\gamma}^\alpha = \sum_{0 \le \gamma_1 \le \beta_1} \sum_{\gamma_2, \dots, \gamma_k} \sum_{\Gamma_{\tilde\alpha}} (-1)^{|\delta|} (I_{\beta-\gamma-\delta}^\alpha - I_{\beta-\gamma-\delta-(1,0)}^\alpha))$$

$$= \sum_{\gamma_2, \dots, \gamma_k} \sum_{\Gamma_{\tilde\alpha}} (-1)^{|\delta|} I_{\beta-\gamma-\delta} |_{\gamma_1=0} = \dots = I_\beta .$$

Q.E.D.

Our next task is to study the Jacobian matrix on the different isotropy subspaces. Let $H < H' \le H_0$ be isotropy subgroups, then

$$E^H = E^{H'} \oplus E' = E^{H_0} \oplus E^\perp \oplus E'.$$

Writing $X = X^0 \oplus X^\perp \oplus X'$ and $f = f^0 \oplus f^\perp \oplus f'$, then from the S^1–equivariance one has $f^\perp(X^0, 0, 0) = 0$ and $f'(X^0, X^\perp, 0) = 0$.

As in Proposition 5.3, the linearization of Df at (X_0, ν^0) will have a diagonal form with terms $f^0_{X^0}, f'_{X^\perp}, f'_{X'}$ which commute with S, where S is the generator of the action of H_0 (apply the arguments of Proposition 5.3 to the couples $(H, H'), (H, H_0), (H', H_0)$).

Since $f'(X^0, X^\perp, X') = f'_{X'}(X^0, X^\perp, 0)X' + o(|X'|)$, then if $I - f'_{X'}(X_0)$ is invertible, it will be also invertible for X^0 close to $X_0, X^\perp$ close to 0 and $X' = 0$, and one may compute $I_H = I_{H'}Index(I - f'_{X'}(X^0))$ by deforming $X' - f'(X^0, X^\perp, X')$ to its linear part (this is a H_0-equivariant deformation) and X' to 0 in the two other components.

If $|H'| = q|H| = qm, |H_0| = r|H'|$ then the modes in E^{H_0} are multiples of rqm, those in $E^\perp$ are multiples of qm but not of rqm and those in E' are multiples of m but not of qm. If S corresponds to $\varphi = 2\pi/rqm$, then S acts on the mode $n_j m$ in E' (thus n_j is not a multiple of q) by $e^{2\pi i n_j/rq}$.

As in Remark 5.2, if X' is a generalized eigenvector of $f'_{X'}$ so is $SX' = \pm X'$. If $SX' = X'$ then X' is in E^{H_0} and $X' = 0$, while if $SX' = -X'$ then the corresponding modes have n_j as odd multiples of $rq/2$. Since n_j cannot be a multiple of q, then r must be odd and q even. Thus $I_H = \eta I_{H'}$ with $\eta = 1$ except if $|H_0|/|H'|$ is odd and $|H'|/|H|$ is even. A priori η depends on $f'_{X'}$, in fact:

Lemma 7.2.

a) *Let* $|H| = P^{\alpha-\beta}, |H'| = P^{\alpha-\beta+\gamma}$, *with* $E^H = E^{H'} \oplus E'$. *If* $I - f'_{X'}$ *is invertible, then* $I^\alpha_\beta = \eta I^\alpha_{\beta-\gamma}$ *with* $\eta = 1$ *unless* P^γ *is even and* $P^{\beta-\gamma}$ *is odd. In this last case, the transition from* $P^\alpha/2$ *to* P^α *is also invertible with linearization* $I - C$ *and* $\eta = index(I - C)$.

b) *If* $\underline{H} < H_j < \overline{H} < H_0, j = 1, 2$ *with* $|H_j| = q_j|\underline{H}|, |\overline{H}| = q_1 q_2 |\underline{H}|, q_1$ *and* q_2 *relatively prime, then if the transition from* $\underline{H}$ *to* H_1 *is invertible, so is the transition from* H_2 *to* $\overline{H}$. $I_{\underline{H}} = \eta I_{H_1}$ *and* $I_{H_2} = \eta I_{\overline{H}}, \eta$ *as above.*

Proof. The remaining part of (a), i.e., the case P^γ even and $P^{\beta-\gamma}$ odd (thus P^β is even), is a consequence of (b). In fact, since the transition from H to H' is invertible, so are the transitions from H to $\underline{H}$, with $|\underline{H}| = |H'|/2$ and from $\underline{H}$ to H' (use the above decomposition for $H < \underline{H} < H' < H_0$).

For the transition from H to $\underline{H}$ one has the couple $(P^\gamma/2, 2P^{\beta-\gamma})$, thus $I_H = I_{\underline{H}}$, and for the transition from $\underline{H}$ to H' one has the couple $(2, P^{\beta-\gamma})$, with $I_{\underline{H}} = \eta I_{H'}$. Since $P^{\beta-\gamma}$ is odd, taking $H_1 = H', |H_2| = P^\alpha/2, \overline{H} = H_0$, one has the result.

For (b), let $E^{\underline{H}} = E^{H_1} \oplus E_1' = E^{H_2} \oplus E_2'$ and $E^{H_j} = E^{\overline{H}} \oplus E_j, j = 1,2$. Since these are orthogonal decompositions (defined on modes) one has $E_1 \oplus E_1' = E_2 \oplus E_2'$. The modes present in E^{H_j} must be multiples of $q_j|\underline{H}|$, hence, for $E^{H_1} \cap E^{H_2}$, the modes must be multiples of $|\overline{H}|$ (q_1 and q_2 are relatively prime) and $E_1 \cap E_2 = \{0\}$. Using again the orthogonality of the mode decomposition, one has that $E^{\underline{H}} = E^{\overline{H}} \oplus E_1 \oplus E_2 \oplus E_3$, where the modes for $E^{\overline{H}}$ are multiples of $q_1 q_2 |\underline{H}|$, those for $E_j, j = 1,2$, are multiples of $q_j|\underline{H}|$ but not of $|\overline{H}|$ and those for E_3 are multiples of $|\underline{H}|$ but not of $|H_j|$.

Decompose X in $E^{\underline{H}}$ as $(\overline{X}, X_1, X_2, X_3)$ and $f^{\underline{H}}$ as $(f^{\overline{H}}, f_1, f_2, f_3)$. Since

$$S(Df^{\underline{H}})(X^0) = (Df^{\underline{H}}(X_0))S$$ one may apply the above argument for the different decompositions of $E^{\underline{H}}$ and E^{H_j} and see that $Df^{\underline{H}}$ is a diagonal application $(D_{\overline{X}}f^{\overline{H}}, D_{X_1}f_1, D_{X_2}f_2, D_{X_3}f_3)$. Thus the transition form $\underline{H}$ to H_1 is invertible if and only if $I - D_{X_2}f_2$ and $I - D_{X_3}f_3$ are invertible. The transition from H_2 to $\overline{H}$ is invertible if and only if $I - D_{X_2}f_2$ is invertible. In this case $I_{\underline{H}} = \text{Index}(I - D_{X_2}f_2)\text{Index}(I - D_{X_3}f_3)I_{H_1}$ and $I_{H_2} = \text{Index}(I - D_{X_2}f_2)I_{\overline{H}}$.

If $|H_0| = q|\overline{H}|, S$ acts on E_3 as $e^{2\pi i k/q_1 q_2 q}$ where k is not a multiple of q_1 or of q_2, hence $2k/q_1 q_2 q$ cannot be an odd integer. Thus $\text{Index}(I - D_{X_3}f_3) = 1$. On the other hand S acts on E_2 as $e^{2\pi i k/q_1 q}$ where k is not a multiple of q_1, hence $k/q_1 q$ will be an odd integer only if q_1 is even and q is odd. Thus $\eta = 1$ unless q_1 is even and q is odd. By repeating the argument in the proof of (a), one may assume $q_1 = 2$ and q odd. Taking then inclusions $\overline{H}/2 < \overline{H} < H_0$ and $\overline{H}/2 < H_0/2 < H_0$ with transitions 2 and q, one will get that $I_{H_2} = I_{\overline{H}/2} = \eta I_{\overline{H}}$ and $I_{H_0/2} = \eta I_{H_0}$ proving the independence of $\eta = \text{Index}(I - D_{X_2}f_2)$ on the particular subgroups $\underline{H}, H_j, \overline{H}$.

Q.E.D.

Definition 7.1. *Given $H < H_0$ with $|H| = P^{\alpha-\beta}, |H_0| = P^\alpha$, we shall say that P^β is a virtual period if for any H' with $H < H' < H_0$, then the map $I - f'_{X'}$ is not invertible. Denote by $\mathcal{M}_\alpha$ the set of all virtual periods and $\{1\}$.*

The justification of the name comes from the fact that if P^β is not in $\mathcal{M}_\alpha$, then $I - f'_{X'}$ is invertible for some H'. Thus if g is a small C^1-perturbation of f, the orbits of $X - g(X, \nu)$

close to (X_0, ν^0) (if any) in E^H will be in fact in $E^{H'}$ with an isotropy type bigger than $P^{\alpha-\beta}$.

Furthermore, in the case of autonomous differential equations, $\mathcal{M}_\alpha$ will coincide with the sets defined in [M.Y] and in [C.M.Y₁] for iterates of maps in $\mathbb{R}^n$. Note that it is enough to verify that none of the transitions from H to H', with $|H'| = p_j|H|$, is invertible, for $\beta_j > 0$ (see Lemma 7.2).

Proposition 7.1.

a) *If P^β and $P^{\beta'}$ are in $\mathcal{M}_\alpha$, so is $P^{\bar\beta}$, with $\bar\beta = (\bar\beta_1, \ldots, \bar\beta_k), \bar\beta_j = max\ (\beta_j, \beta'_j)$, the least common multiple.*

b) *If P^β is not in $\mathcal{M}_\alpha$, then $d^\alpha_{\alpha-\beta} = 0$, except if $P^\beta = 2P^{\tilde\beta}$, with $P^{\tilde\beta}$ odd and in $\mathcal{M}_\alpha$, in which case $d^\alpha_{\alpha-p} = (\eta - 1)d^\alpha_{\alpha-\tilde\beta}/2, \eta$ as in Lemma 7.2.*

c) $I_\beta = \sum_{\delta \in N_\beta} P^\delta d^\alpha_{\alpha-\delta} - (1-\eta) \sum_{\delta \in N'_\beta} P^\delta d^\alpha_{\alpha-\delta}$ *where* $N_\beta = \{\delta, 0 \le \delta \le \beta, P^\beta \in \mathcal{M}_\alpha\}$
and $N'_\beta = \{\delta, 0 \le \delta \le \beta, P^\delta$ *odd in* $\mathcal{M}_\alpha, 2P^\delta \notin \mathcal{M}_\alpha, P^\beta$ *is even* $\}$.

Proof. (a). Consider the transition from H to H_1, with $|H| = P^{\alpha-\bar\beta}, |H_1| = p_1|H|$ (if $\beta_1 > 0$). Since $\bar\beta_1 = max(\beta_1, \beta'_1)$, assume $\bar\beta_1 = \beta_1$. Let H_2 with $|H_2| = P^{\alpha-\beta} = q_2|H|$, where q_2 has no factor p_1. Let $\bar H$ with $|\bar H| = p_1|H_2|$. Since $P^{\alpha-\beta}$ is in $\mathcal{M}_\alpha$, the transition from H_2 to $\bar H$ is not invertible, thus, from Lemma 7.2, nor is the transition from H to H_1.

(b). Assume that the transition from H to H_1 is invertible, with $|H_1| = p_1|H|$. As in Lemma 7.1, write

$$P^\beta d^\alpha_{\alpha-\beta} = \sum_{\Gamma_{\tilde\alpha}} (-1)^{|\delta|}(I^\alpha_{\beta-\delta} - I^\alpha_{\beta-\delta-(1,0)}) .$$

Let $|H_2| = P^{\alpha-\beta+\delta}, |\bar H| = p_1|H_2|$. From the definition of $\Gamma_{\tilde\alpha}, P^\delta$ and p_1 are relatively prime. Thus, from Lemma 7.2, $I^\alpha_\beta = \eta I^\alpha_{\beta-(1,0)}, I^\alpha_{\beta-\delta} = \eta I^\alpha_{\beta-\delta-(1,0)}, \eta = 1$ except if p_1 is $2, \beta_1 = 1$. Hence $d^\alpha_{\alpha-\beta} = 0$ unless the only invertible transition is for $p_1 = 2$ with $P^\beta = 2P^{\tilde\beta}$ and $P^{\tilde\beta}$ odd (if P^β is even use Lemma 7.2 (a)). In this case $P^\beta d^\alpha_{\alpha-\beta} = (\eta-1)\sum_{\Gamma_{\tilde\alpha}} (-1)^{|\delta|} I^\alpha_{\beta-\delta} = (\eta-1)P^{\tilde\beta} d^\alpha_{\alpha-\tilde\beta}$.

(c). Is clear from (b) and Lemma 7.1

Q.E.D.

Remark 7.1. For an autonomous differential equation, with an isolated periodic solution taken at $\nu = \nu_0/P^\alpha$ (ν_0 the maximal frequency), one has to consider the kernel of

$I - \Psi(T_0)^{|H_0|/|H|}$. (see 5.4). By looking in parallel at the maps $I - \Psi(T_0)^k$ and $I - f'_{X'}$ for the subgroups H', with $|H_0| = P^\beta |H|, |H'| = p_j |H|$, it is easy to see that

$$P^\beta \ is \ a \ virtual \ period \ in \ \mathcal{M}_\alpha \ if \ and \ only \ if \ there \ is \ a \ Y \ in \ \mathbb{R}^M \ such \ that$$
$$\Psi(T_0)^{P^\gamma} Y \neq Y \ for \ \gamma < \beta \ and \ \Psi(T_0)^{P^\beta} Y = Y.$$

Furthermore, by perturbing the system to a generic situation, one has that I_β^α is the index of the Poincaré section for the P^β return map and, thus, $I_\beta^\alpha = I_\beta^\beta \equiv I_{P^\beta}$. From the fact that $\Gamma_\alpha = \Gamma_\beta$ and from Lemma 7.1, one has that $d_{\alpha-\beta}^\alpha = d_0^\beta \equiv d_{P^\beta}$. In this case η, defined only if 2 is not a virtual period, is $(-1)^{\sigma_-}$ (σ_- as in Proposition 5.6). Then the relation in Proposition 7.1 (c) reads as follows.

For any positive integer m

$$I_m = \sum_{\substack{k \in \mathcal{M} \\ k|m}} k d_k - (1 - \eta) \sum_{\substack{k \text{ odd} \in \mathcal{M} \\ 2k \notin \mathcal{M}, 2k|m}} k d_k$$

where $\mathcal{M}$ is the set of all virtual periods.

In [C.M.Y$_1$] one has the formula

$$I_m = \sum_{\substack{k \in \mathcal{M} \\ k|m}} k c_k \qquad if \ \sigma_- \ is \ even$$

$$I_m = \sum_{\substack{2k \in \mathcal{M} \\ 2k|m}} 2k c_{2k} - \sum_{\substack{2k+1 \in \mathcal{M} \\ 2k+1|m}} (2k+1)(-1)^{m/(2k+1)} c_{2k+1} \ if \ \sigma_- \ is \ odd .$$

Then, if σ_- is even, $c_k = d_k$ (if $2 \in \mathcal{M}$ the second sum is empty by Proposition 7.1 (a)). If σ_- is odd then $d_{2k+1} = c_{2k+1}, d_{4k} = c_{4k}, d_{4k+2} + d_{2k+1} = c_{4k+2}$ (recall that if 2 and k are in $\mathcal{M}$, then $2k$ is in $\mathcal{M}$ if k is odd). (These identities may be obtained by direct substitution or by looking at the different parities for m, σ_- and η). Thus the two approaches coincide.

Finally if $k_1, \ldots, k_s$ are the elements of $\mathcal{M}$, let M be a multiple of $k_1, \ldots, k_s$ and one has:

$$\sum_1^M I_m = \sum k_j d_{k_j} M/k_j - (1 - \eta) \sum_{\substack{k_j \text{ odd} \\ 2k_j \notin \mathcal{M}}} k_j d_{k_j} M/(2k_j)$$

where M/k_j is the number of m's which are multiple of k_j and $M/(2k_j)$ the number of m's which are multiple of $2k_j$. Thus

$$\frac{1}{M}(\sum_1^M I_m) = \sum_{k_j \in M} d_{k_j} - (1-\eta)/2(\sum_{\substack{k_j \text{ odd} \in M \\ 2k_j \notin M}} d_{k_j})$$

recovering also the result of [C.M.Y$_1$].

7.2. The orbit index.

Autonomous differential equations have the special feature that if $X_0(t)$ is a T–periodic solution then it is also a kT–periodic solution. In terms of Fourier series for $\nu dX/d\tau - g(X(\tau))$, let P_k be the isomorphism from $E \times \mathbb{R}$, onto $E^k \times \mathbb{R}$ with $E = H^1(S^1)^M$, E^k is the space of all modes which are multiples of k, defined as follows

$$P_k(X,\nu) = P_k(X_0, X_1, X_2, \dots, \nu) = (X_0, 0, \dots, 0, X_1, 0, \dots, 0, X_1, \dots, \nu/k)$$

i.e., if $\tilde{X}_m$ is the m'th mode of P_k, then $\tilde{X}_m = 0$ if m is not a multiple of k and $\tilde{X}_{kn} = X_n$.

From a change of variable in the integral defining $g_m(X)$ and from the equivariance, one sees that $g_m(P_k(X)) = 0$ if $m \neq kn$ and $g_{kn}(P_k(X)) = g_n(X)$. Thus if

$$G(X,\nu) = (g_0(X), g_1(X)/i\nu, \dots, g_n(X)/in\nu, \dots,)$$

then

$$G(P_k(X,\nu)) = P_k G(X)$$

and if X is a fixed point of $G(X,\nu)$, then $P_k(X,\nu)$ will also be a fixed point of G, but in E^k. Note that $\|P_k X\|_{L^2} = \|X\|_{L^2}$ but

$$\|P_k X\|_{H^1} = (\|P_k X\|^2 + \|(P_k X)'\|^2)^{1/2} = (\|X\|^2 + k^2 \|X'\|^2)^{1/2} .$$

As in 5.6 and Proposition 5.9, let ω be an open bounded subset of $\mathbb{R}^M$ such that the system of differential equations has no periodic solutions on its boundary. Let

$$\Omega = \{(X,\nu) \text{ in } H^1(S^1)^M \times \mathbb{R}, X(\tau) \in \omega \text{ for all } \tau \text{ in}[0, 2\pi], \nu > 0\}.$$

Ω is open and unbounded and there are two constants N_1 and N_2 such that any X, fixed point of $G(X, \nu)$, has $\|X\|_1 \leq N_1 + N_2/\nu$.

If $X(\tau)$ is a 2π–periodic solution and g is Lipschitz–continuous on $\bar{\omega}$, then $\nu^2|X'(\tau) - X'(0)|^2 \leq L^2|X(\tau) - X(0)|^2$. Then, using the periodicity of $X(\tau)$, one gets

$$\nu^2\|X'\|^2 \leq \nu^2(\|X'\|^2 + |X'(0)|^2) \leq 2\pi^2 L^2\|X'\|^2 \ .$$

Thus, for a non–stationary solution, $\nu \leq \pi\sqrt{2}\,L$. (In fact, one may improve the bound to $\nu \leq L$, from a result of Lasota and Yorke).

Furthermore if there are no stationary solutions in $\bar{\omega}$, thus $|g(X)| \geq K > 0$, then $\nu\|X'\| \geq K$.

The above considerations motivate the attempt to define a S^1–degree for ω, that is for Ω through a limiting process. More precisely, in the generic case of a map with only hyperbolic orbits, then each orbit X^0, as well as its replicae $P_k(X^0)$, is isolated in Ω. At $P_k(X^0, \nu^0) = (P_k(X^0), \nu^0/k)$, with ν^0 the maximal frequency, the S^1-index is equal to $(0, d_1^k, d_2^k, \ldots, d_k^k, 0, \ldots)$ with $d_k^k = (-1)^{\sigma+} = I_1, d_j^k = 0$ except if k is even, in which case $d_{k/2}^k = (I_2 - I_1)/2((-1)^{\sigma+ +\sigma-} - (-1)^{\sigma+})/2$ (see Proposition 5.6), taking into account the change of notation.

In this generic case the set of solutions is finite, thus the set of periods is bounded. There is an infinite number of R_n, R_n tending to ∞, such that on $\Omega_n = \{(X, \nu), \|X\|_1 < R_n\}$, the set of zeros is compact (contained in the region defined by the above bounds). Thus $\deg_{S^1}(X - G(X, \nu), \Omega_n)$ is defined. If $X^0, P_1 X^0, \ldots, P_k X^0$ are in Ω_n, then the contribution of X^0 and these replicae to the S^1–index will be $(0, I_1 + I_2)/2, \ldots, (I_1 + I_2)/2, I_1, \ldots, I_1)$, with $[k/2]$ terms $(I_1 + I_2)/2$.

When n goes to ∞, then the limiting contribution will be $(0, \varphi(X^0), \varphi(X^0), \ldots)$, where $\varphi(X^0) = \lim(\sum_1^M I_m)/M$ is the orbit index defined in [A.M.Y], [M.Y] and [C.M.Y$_1$]: in this generic case, as seen in Proposition 5.6:

$$\varphi(X^0) = (I_1 + I_2)/2.$$

For an isolated orbit, the above argument will lead to a limiting S^1–degree of the form $(0, \varphi(X^0), \varphi(X^0), \ldots)$ as it follows from Proposition 7.1 and Remark 7.1.

Note that the fact that there is an infinite number of non–zero components is not surprising in view of the non–boundedness of the set of zeros in Ω and the lack of compactness of the map when ν tends to 0.

For the general case, the approach used in [M.Y] (and in preceding work) is to perturb the map G to the generic case of hyperbolic orbits. This requires a strict control on the periods of the perturbed system, that is, as explained in [M.Y] and in §7.1, a bound on the set of virtual periods for the periodic solutions of $dX/dt = g(X)$. It is then clear that the arguments used in [M.Y] would equally work here. One has thus:

Proposition 7.2. *The orbit index, as defined in [M.Y], is equal to all components of* $\deg_{S^1}(X - G(X, \nu), \Omega)$.

Virtual periods have the unpleasant property of being not continuous with respect to perturbations. Thus, in order to verify that a family of differential equations have their virtual periods bounded on a set ω, one needs either to know all solutions (in which case one could have doubts about the role of the index) or one needs to make the apriori hypothesis of that boundedness (as done in the applications to global Hopf bifurcation and continuation) and then one has to be very careful to check the hypothesis for any deformation used in order to compute this index.

An alternate route would then be to avoid the generic perturbation and see under which hypothesis a limiting process on sets Ω_n, with Ω_n tending to Ω, would be stable. A necessary hypothesis (in order to have a bounded subset of Ω), will be to assume that

if C is any component of zeros of $X - G(X, \nu)$ in Ω, then C is bounded.

This implies, for the corresponding component of periodic solutions in ω, not only that it has no stationary solution on it and that the minimal periods on it are bounded, but also that

C has no "period multiplying loop" in the sense that $C \cap P_k(C) = \emptyset$, for all $k \geq 2$.

In fact if (X, ν) and $P_k(X, \nu)$ belong to C, then $P_k(X, \nu)$ and $P_{k^2}(X, \nu)$ belong to $P_k(C)$. But, since C is a connected component and $P_k(C)$ is connected, then $P_k(C)$ is contained in C. One gets then that $P_{k^n}(X, \nu)$ is in C for all n's. Since $\|P_{k^n}(X, \nu)\|_1$ goes to ∞, C is unbounded. In fact one has

Lemma 7.3. *If the minimal periods on ω are bounded and there are no stationary solutions in $\overline{\omega}$, then a component in Ω is unbounded if and only if some component has a period multiplying loop.*

Proof. Let us define an orbit (X, ν) as a *jump orbit* if there is a sequence (X_n, ν_n), with isotropy $\{e\}$ (i.e., with maximal frequency) such that (X_n, ν_n) converges, in $H^1 \times \mathbb{R}$, to $P_k(X, \nu)$. We shall say that k is the *jump*.

Notice first that the set of jumps is a subset of the set of virtual periods.

But in the conditions of the Lemma the set of jumps is finite: for a jump one has $\| \overset{\bullet}{X}_n \| \simeq \| P_k(X)^\bullet \| = k \| \overset{\bullet}{X} \|$, but in ω, $\| \overset{\bullet}{X}_n \|$ is bounded from above and $\| \overset{\bullet}{X} \|$ is bounded from below.

Let J be the set of jump orbits in Ω. Then J is closed (hence compact since J is bounded and by the compactness of the map $G(X, \nu)$). In fact if (X_n, ν_n) tends to (X, ν), with $P_{k_n}(X_n, \nu_n)$ close to $(\overline{X}_n, \overline{\nu}_n)$, one may choose a subsequence with $k_n = k$ (finite number of possible values for k_n). One may then choose $(\overline{X}_n, \overline{\nu}_n)$ tending to $P_k(X, \nu)$. (Note that we are not requiring that X has maximal frequency, thus it could come itself from another jump; a finite number of them because of the bounds on the maximal frequency).

A last preliminary fact is the following: let $X(\tau)$ be a periodic solution in ω, corresponding to (X, ν) with ν maximal. Let K be the least common multiple of the set of jumps. Given a neighborhood N of $P_K(X, \nu)$ in $E \times \mathbb{R}$, then there is a neighborhood in $\mathbb{R}^M$ of the orbit $\{X(\tau)\}$ such that any periodic solution of the equation contained in the neighborhood has a replica in N. In fact, if this is not the case there would be periodic solutions $X_n(\tau)$ with orbits converging to the orbit of $X(\tau)$ and $\| P_k(X_n, \nu_n) - P_K(X, \nu) \|_1 \geq \varepsilon$ for all divisors k of K (ν_n is maximal here). By the compactness, (X_n, ν_n) tends to $(\overline{X}, \overline{\nu})$ which, by uniqueness of the initial value problem, is $P_k(X, \nu)$ for some k. Thus k is a jump and must divide K. Then $P_{K/k}(X_n, \nu_n)$ tends to $P_K(X, \nu)$.

Let then $\mathcal{C}$ be an unbounded connected component of zeros in Ω.

Since $P_k(\mathcal{C})$ is connected and unbounded, we may as well assume that $\mathcal{C}$ is contained in $E^K \times \mathbb{R}$, where E^K corresponds to all modes multiple of K.

Since the set of maximal frequency solutions is bounded, one has on $\mathcal{C}$ an infinite number of points which are of the form $P_{n_j k_j}(X, \nu)$, where k_j is a jump, and points that are limits of sequences of the form $P_{n_j}(X_1, \nu_1)$.

Furthermore, up to a common factor, the n_j's have to come from the finite number of jumps, that is, if $k_1, \ldots, k_s$ are the prime factors of the jumps, then $n_j = A k_1^{\alpha_{1j}} \ldots k_s^{\alpha_{sj}}, \alpha_j \geq 0$. In fact one may choose the n_j's such that n_{j+1} is a multiple of n_j: if for the set of multi-integers $(\alpha_{1j}, \ldots, \alpha_{sj})$, one has that $\{\alpha_{1j}, \ldots, \alpha_{kj}\}$ are bounded and the others are not, then one may choose a subsequence such that $\alpha_{1j}, \ldots, \alpha_{kj}$ are constant. All the other indices must be unbounded (one may have to take again a subsequence) then one may choose a set of indices which are ordered in strictly increasing sequences. Furthermore one may choose the sequence such that the jump k_j is constant.

If $\mathcal{C} \cap P_m(\mathcal{C}) = \emptyset$ for all $m > 1$, then (X_{n_j}, ν_{n_j}) with $P_{n_j k}(X_{n_j}, \nu_{n_j})$ in $\mathcal{C}$, are jump orbits which are all different (if two are equal for n_1 and n_2, then $P_{n_1 k}(X, \nu)$ and $P_{n_2 k}(X, \nu)$ are two points of $\mathcal{C}$ and, since n_2 is a multiple of n_1, then $\mathcal{C} \cap P_{n_2/n_1}(\mathcal{C}) \neq \emptyset$).

As before the jump orbits (X_{n_j}, ν_{n_j}) accumulate at $(\overline{X}, \overline{\nu})$. Take a small neighborhood N of $P_K(\overline{X}, \overline{\nu})$ which reproduces the neighborhood in $\mathbb{R}^M$ of $\overline{X}(\tau)$. Since the image in $\mathbb{R}^M$ of $\mathcal{C}$ is connected and compact, each orbit $\{X_{n_j}(\tau)\}$ belongs to a local component which is reproduced as C_{n_j} in N.

If $C_{n_1} \cap C_{n_2} \neq \emptyset$, then $\mathcal{C}$ would have a loop, since n_1 is a multiple of n_2 and K divides n_j : $P_{n_j k}(X_{n_j} \nu_{n_j})$ belongs to a $P_{n_j k/K}(C_{n_j})$ and $P_{n_1 k}(X_{n_2}, \nu_{n_2})$ is connected to $P_{n_1 k}(X_{n_1}, \nu_{n_1})$. The same situation prevails on a small neighborhood of $P_{n_1 k/K}(N)$ (no new solution may be H^1-close to $P_{n_1 k/K}(N)$ since then it would be close in $\mathbb{R}^M$ to the orbit $\overline{X}(\tau)$ and hence come as a replica from N). But there $P_{n_1 k/K}(C_{n_1})$ is just $\mathcal{C}$ and joins $P_{n_1 k}(X_{n_1}, \nu_{n_1})$ to $P_{n_2 k}(X_{n_2}, \nu_{n_2})$.

A covering argument of this piece of $\mathcal{C}$ and the above selection of n_j's will show that $P_{n_1 k/K}(C_{n_j})$ is part of the replica of the piece of $\mathcal{C}$ between $P_{n_2 k}(X_{n_2}, \nu_{n_2})$ and $P_{n_3 k}(X_{n_3}, \nu_{n_3})$, with $P_{n_2 k}(X_{n_3}, \nu_{n_3})$ on it. These connected components must accumulate to a connected component (use the separation result for components of compact sets) from $P_{n_1 k}(\overline{X}, \overline{\nu})$ to $P_{n_2 k}(\overline{X}, \overline{\nu})$ obtaining thus a component with a loop (n_2 is a multiple of n_1).

Thus, under these hypotheses, the image of $\mathcal{C}$ in $\mathbb{R}^M$ enters the neighborhood of the orbit $X(\tau)$ an infinite number of times, has a jump there and recovers this jump of frequency outside the neighborhood. At the limit, the image of $\mathcal{C}$ has a loop going through $X(\tau)$. The passage to N is made so that one recovers the lack of continuity in the frequency and finally one uncoils the spiral along $\mathcal{C}$.   Q.E.D.

Assume thus that all components in Ω are bounded. Let

$$\Omega_R = \{(X, \nu) \in \Omega, \|x\|_1 < R\} = \Omega \cap B_R.$$

Consider the set S_R of zeros of $X - G(X, \nu)$ which belong to connected components in Ω_R or emanating from $\overline{\Omega}_R$.

S_R is in fact bounded: if not one would have a sequence of zeros (X_n, ν_n) in $\overline{\Omega}_R$ and of connected components C_n joining (X_n, ν_n) to a point with H^1–norm bigger than nR. By the compactness of the set of zeros in Ω_R, one would have a limit point $(\overline{X}, \overline{\nu})$ and a limit component $\overline{C}$ which cannot be bounded: if it were bounded construct a neighborhood of $\overline{C}$ with no zeros on its boundary. But for n large enough (X_n, ν_n) should be in this neighborhood and thus C_n would intersect its boundary.

One may then construct an open bounded subset V_R of Ω such that $S_R \subset V_R$ and V_R has no zeros on its boundary. Since S_R is not connected one cannot apply directly the separation result for connected components, but if one defines as $\tilde{S}_R = \{f(\tau, \|X\|_1)(X, \nu), (X, \nu) \in S_R, f(\tau, \|X\|_1) \text{ is } \tau \text{ if } \|X\|_1 \leq R, \text{ and } 1 \text{ if } \|X\|_1 > R\}$ (i.e., one joins linearly all points in $S_R \cap B_R$ to the origin), then $\tilde{S}_R$ is compact and connected (by the definition of S_R). One will find a neighborhood of $\tilde{S}_R$ with the right properties. Call V_R its intersection with Ω. Since all zeros in Ω_R are in S_R, the construction of $\tilde{S}_R$ does not modify the situation there. We have thus shown:

Proposition 7.3. *If all periodic solutions in $\overline{\omega}$ are in ω and have their minimal period bounded from above, if there are no stationary solutions in $\overline{\omega}$ and there is no period multiplying loop, then for every R, there is a bounded open set V_R in Ω, containing all solutions with H^1–norm less than R, with no zeros on its boundary, i.e., $\deg_{S^1}(X - G(X, \nu); V_R)$ is well defined.*

It is clear from the construction that if $R' > R$, then one may construct $V_{R'} \supset V_R$. For a sufficiently small C°-perturbation of the field $g(X)$, the corresponding perturbation of G on V_R will be H^1-close and the two S^1-degrees will be equal. This will be true in particular for the generic perturbations. However the size of the perturbation depends on R (through V_R).

Thus if R_n tends to ∞, the limit of $\deg_{S^1}(X - G(X,\nu), V_{R_n})$ is not guaranteed.

Note that if this degree is $(0, d_1^n, d_2^n, \ldots)$ it is enough to have a limit d for d_1^n in order to have a limit for d_j^n and, in fact, this limit is d: for n large enough $d_1^n = d$ and the ulterior contributions by generic perturbations to d_1^n are zero. From Proposition 7.3 these contributions to d_j^n will also be zero but for larger values of n. For these perturbations, the previous contributions will be equal for $d_1, d_2 \ldots, d_N$ with N going to infinity with n.

In general the limit of the S^1-degrees will exist if there are no perturbations with periodic solutions of arbitrarily large period (this is implied by a bound on the sets of virtual periods, and it is probably equivalent) or if these periodic orbits have a total contribution to the orbit index equal to zero.

Another interesting point would be to know if an isolated and bounded loop of periodic solutions (with bounded virtual periods) has an orbit index, as defined in [A.M.Y] and [M.Y], equal to zero. An indication of this is to take the example, in these papers, of a period doubling loop for an equation $dX/dt = f(X, \lambda)$. Then one may close the loop constructed there by a symmetric loop in the other direction. Then one completes the differential system with the equation $d\lambda/dt = 0$. The orbit index for the couple is zero, since one may perturb the second equation to $d\lambda/dt = \varepsilon$, which has no periodic solution and the perturbation doesn't change the virtual periods.

Remark 7.2. For the bifurcation situation of Proposition 6.4, one may compute the orbit index for the map $(\varepsilon^2 - \| \dot{X} \|^2, X - f(X, \nu, \lambda))$ as limit of S^1-indices on sets of the form $\{\|X - X_0\|_1 < 2\varepsilon, |\lambda - \lambda_0| < \rho, \nu_n < \nu < \nu_0\}$ with ν_n tending to zero (one encounters an infinite number of components with no contribution to the orbit index).

The orbit index will then be $\eta_0 \sum \sigma_j$, with $\eta_0 = \text{Index}(g(X, 0); 0)$ as in Proposition 6.4 and $\sum \sigma_j$ is the total crossing number.

The same result will be true when the complementing function is $\varepsilon^2 - |X(\tau) - X_0|^2_{C^0}$, giving the following global bifurcation result:

if $|X(\tau)|_{C^0}$ is bounded and the set of virtual periods is bounded on the branch (so that one may use the usual argument) one may deform ε^2 to R^2, for large R, obtaining that the local orbit index must be zero.

Thus if the total crossing number is non–zero one has global bifurcation (with virtual periods).

Note that the above deformation does not alter the virtual periods. However one may have $\sum \sigma_j = 0$ but several of the σ_j's non–zero, with global bifurcation in the sense of [$\mathbf{I_0}$]. Since the orbit index eliminates the frequency, the usual trick of choosing a frequency cannot be applied. This discrepancy may be an indication that the orbit index is not as complete as the S^1–index (in the sense of finding perturbations without zeros on simple sets if the index is zero). A first study of the differences between the results obtained by these two methods can be done, in the case of the Hopf bifurcation, by analyzing the results of [$\mathbf{A.Y}$]. Then the orbit index could be used as a path following device for the generic case (as done in [$\mathbf{A.M.Y}$], [$\mathbf{M.Y}$] and related works).

APPENDIX

ADDITIVITY UP TO ONE SUSPENSION

One of the slightly irritating points of the degree theory defined in [**I.M.V**] is that

$$\Sigma_0 \deg_\Gamma(f; \Omega_1 \cup \Omega_2) = \Sigma_0 \deg_\Gamma(f; \Omega_1) + \Sigma_0 \deg_\Gamma(f; \Omega_2)$$
$$\text{if } \bar{\Omega}_1 \cap \bar{\Omega}_2 = \emptyset.$$

Thus, the degree is fully additive only if the suspension Σ_0 is a monomorphism. A doubt was left about the fact that this suspension was really necessary.

F. Romero Ruiz del Portal has settled this point by constructing an example in [**R**] for which the equality

$$\deg(f; \Omega_1 \cup \Omega_2) = \deg(f; \Omega_1) + \deg(f; \Omega_2)$$

is not true.

In his doctoral thesis, F. Romero Ruiz del Portal, defines a degree theory, for non-equivariant maps, based on framed cobordism instead of our approach. His theory has the same properties as ours and, from the universality property and from the Pontryagin-Thom isomorphism, they coincide. The equivariant case is not treated and would require a strong equivariant Sard's lemma in order to approximate the maps by a C^∞ equivariant map for which one could use the ideas of framed equivariant cobordism.

In this appendix, we shall give an example similar, although a little different, to the one presented in [**R**] (there $\deg(f; \Omega_1 \cup \Omega_2) = 2$, $\deg(f; \Omega_j) = 0$, while for our example $\deg(f; \Omega_1 \cup \Omega_2) = 0$ and $\deg(f; \Omega_j) = 1$).

In $\mathbb{R}^3$, consider the map defined as

$$f(x_1, x_2, x_3) \equiv f_1 + if_2 \equiv (x_1^2 + x_2^2 - 1 + ix_3)((x_1 - 1)^2 + x_3^2 - 1 + ix_2)$$

(the complex notation is used for brevity only).

The zeros of f are the two linked circles $+S_1 = \{x_1^2 + x_2^2 = 1, x_3 = 0\}$ and $S_2 = \{(x_1 - 1)^2 + x_3^2 = 1, x_2 = 0\}$.

Take $B = \{(x_1, x_2, x_3)/x_1^2 + x_2^2 + x_3^2 < 4\}$ and Ω_j small disjoint tubular neighborhoods of S_j.

By the excision property, $\deg(f; B) = \deg(f; \Omega_1 \cup \Omega_2)$. Since the zeros of f are in B, one may use the same function f in order to compute $\deg(f; B) = [2t + 2\phi(X) - 1,$ $f(X)]_{I \times B_R} = [2t - 1, f(X)]_{I \times B}$ by scaling B_R to B. Hence $\deg(f; B) = \Sigma_0[f]$, from the suspension property given in the Introduction. Here $[f]$ is the homotopy class of f as a map from $\partial B \cong S^2$ into $\mathbb{R}^2 \backslash \{0\}$, i.e., an element of $\Pi_2(S^1) = 0$. Thus, $\deg(f; B) = 0$.

On the other hand, in Ω_1, one may perform the deformation:

$$(x_1^2 + x_2^2 - ix_3)(\tau((x_1 - 1)^2 + x_3^2 - 1) - (1 - \tau)x_1 + ix_2).$$

The zeros of the deformation in $\partial\Omega_1$ (thus the first factor is non-zero) would be for $x_2 = 0$ hence x_1 would be close to ± 1 and x_3 close to 0. For (x_1, x_3) close to $(1,0)$ the deformed term is negative, while, for (x_1, x_3) close to $(-1, 0)$, this term is positive.

Thus, $\deg(f; \Omega_1) = \deg((x_1 - ix_2)(x_1^2 + x_2^2 - 1 + ix_3); \Omega_1)$ (the change (x_1, x_2) into $(-x_1, -x_2)$ is orientation preserving). Call $z = x_1 + ix_2$. Then $\deg(\bar{z}(|z|^2 - 1 + ix_3); \Omega_1)$ is the class of $(2t + 2\phi(X) - 1, \bar{z}(|z|^2 - 1 + ix_3))$ in $\Pi_3(S^2) \cong \mathbb{Z}$, where $\phi(X) = 0$ in Ω_1, $\phi(X) = 1$ outside a neighborhood of Ω_1. One may deform linearly $\phi(X)$ to $(|z|^2 - 1)^2$ (in fact at $z = 0, \phi(0, x_3) = 1$). One may also take the deformation $(|z|^2 - 1)(1 + \tau)(1 - \tau(2t - 1)) + ix_3$, since $1 - \tau(2t - 1) \geq 0$ and it is 0 only for $\tau = t = 1$, then $2t - 1 = 1$. The resulting map is:

$$(2t - 1 + 2(|z|^2 - 1)^2, \bar{z}(4(|z|^2 - 1)(1 - t) + ix_3)).$$

On the first component and on the term $2(1 - t)(|z|^2 - 1)$, perform the deformation:

$$\begin{pmatrix} \tau & -(1 - \tau) + 2\tau(|z|^2 - 1) \\ 1 - \tau - 2\tau(|z|^2 - 1) & 2\tau \end{pmatrix} \begin{pmatrix} 2t - 1 \\ |z|^2 - 1 \end{pmatrix}.$$

Then $\deg(f;\Omega_1)\eta = [1 - |z|^2, \bar{z}(2t - 1 + ix_3)] = \eta$

where η is the Hopf map. Thus, $\deg(f;\Omega_1) = 1$.

In order to compute $\deg(f;\Omega_2)$, make the deformation

$$(\tau(x_1^2 + x_2^2 - 1) + (1 - \tau)(x_1 - 1) + ix_3)((x_1 - 1)^2 + x_3^2 - 1 + ix_2).$$

On $\partial\Omega_2$, an eventual zero would be for $x_3 = 0$, (x_1, x_2) close to $(2, 0)$ or to $(0, 0)$. In the first case the deformed term is positive, while in the second case it is negative.

The resulting map $(x_1 - 1 + ix_3)((x_1 - 1)^2 + x_3^2 - 1 + ix_2)$ can be written as

$(y_1 - iy_2)(y_1^2 + y_2^2 - 1 + iy_3)$ under the change of variables $y_1 = x_1 - 1, y_2 = -x_3, y_3 = x_2$, with positive Jacobian, and Ω_2 is sent in Ω_1. Then one has $\deg(f;\Omega_2) = 1$.

Thus $\deg(f;\Omega_1 \cup \Omega_2) = 0 \neq \deg(f,\Omega_1) + \deg(f;\Omega_2) = 2$.

Clearly, when one suspends, then the equality holds.

This is why we have restricted our degree, for the case $l = 2$, to maps which have $d_0 = 0$.

REFERENCES

[**A.M.Y**] K.T. Alligood, J. Mallet–Paret and J.A. Yorke, *An index for the global continuation of relatively isolated sets of periodic orbits*, Lect. Notes in Math., 1007, Springer–Verlag (1983), 1–21.

[**A.Y**] K.T. Alligood and J.A. Yorke, *Hopf bifurcation: the appearance of critical periods in cases of resonance*, J. Diff. Eq., 64 (1986), 375–394.

[**B**] G.E. Bredon, *Introduction to compact transformation groups*, Academic Press, New–York 1972.

[**C.M.Y**] S.N. Chow, J. Mallet–Paret and J.A. Yorke, *Global Hopf bifurcation from a multiple eigenvalue*, Nonlinear Anal., M.T.A., 2 (1978), 755–763.

[**C.M.Y$_1$**] S.N. Chow, J. Mallet–Paret and J.A. Yorke, *A periodic orbit index which is a bifurcation invariant*, Lect. Notes in Math., 1007, (1983), 109–131.

[**C.M.Z**] A. Capietto, J. Mawhin and F. Zanolin, *Continuation theorems for periodic perturbations of autonomous systems.* To appear in Trans. Amer. Math. Soc.

[**Da**] E.N. Dancer, *A new degree for S^1-invariant gradient mappings and applications*, Ann. Inst. Henri Poincaré, Anal. Non Linéaire, 2 (1985), 329–370.

[**D.T**] E.N. Dancer and J.F. Toland, *Degree theory for orbits of prescribed period of flows with a first integral*, Proc. London Math. Soc. 3, 60 (1990), 549–580.

[**D.G.J.M**] G. Dylawerski, K. Geba, J. Jodel and W. Marzantowicz, *An S^1-equivariant degree and the Fuller index*, Univ. of Gdansk, Preprint no. 64, 1987.

[**F**] F.B. Fuller, *An index of fixed point type for periodic orbits*, Amer. J. Math, 89 (1967), 133–148.

[**F.R**] E.R. Fadell and P.H. Rabinowitz, *Generalized Cohomological index theories for Lie group actions with an Application to Bifurcation Questions for Hamiltonian Systems*, Inventiones Math. 45, 139-174, 1978.

[**G.L.R**] I. Gohberg, P. Lancaster L. Rodman, *Matrices and indefinite scalar products*. Birkhauser Verlag. 1983.

[$\mathbf{I}_0$] J. Ize, *Obstruction theory and multiparameter Hopf bifurcation*, Trans. Amer. Math. Soc., 291, (1985), 383–435.

[$\mathbf{I}_1$] J. Ize, *Necessary and sufficient conditions for multiparameter bifurcation*, Rocky Mountain J. of Math., 18 (1988), 305–337.

[**I.M.P.V**] J. Ize, I. Massabó, J. Pejsachowicz and A. Vignoli, *Structure and dimension of global branches of solutions to multiparameter nonlinear equations*, Trans. Amer. Math. Soc. 291, (1985), 383-435.

[$\mathbf{I.M.V}_0$] J. Ize, I. Massabó and A. Vignoli, *Global results on continuation and bifurcation for equivariant maps*, Nonlinear Functional Analysis and its Applications, (Proc. NATO Advanced Study Institute, Maratea, Italy, 1985), Mathematical and Physical Sciences, Vol. 173, Reidel Publishing Co., Dordrecht, Holland, 1986, 74-111.

[**I.M.V**] J. Ize, I. Massabó and A. Vignoli, *Degree Theory for equivariant maps*, I, Trans. Amer. Math. Soc., 315 (1989), 433–510.

[**M.Y**] J. Mallet–Paret and J.A. Yorke, *Snakes: oriented families of periodic orbits, their sources, sinks and continuation*, J. Diff. Eq., 43 (1982), 419–450.

[**Na**] U. Namboodiri, *Equivariant vector fields on spheres*, Trans. Amer. Math. Soc., 278 (1983), 431–460.

[**R**] F. Romero Ruiz del Portal, *Teoría del Grado Topológico Generalizado y Aplicaciones*. Tesis Doctoral, Universidad Complutense de Madrid, Marzo 1990.

[**t.D**] T. tom Dieck, *Transformation groups and representation theory*, Lect. Notes in Math. 766, Springer Verlag, New York, 1979.

[**V**] A. Vanderbauwhede, *Local bifurcation and symmetry*, Research notes in Math.
Pitman, 1982.

[**W**] G.W. Whitehead, *Elements of Homotopy Theory*, Graduate Texts in Mathematics, Springer–Verlag, 1978.

*JORGE IZE. UNAM-IIMAS, Apdo. Postal 20-726, Admón. 20, Deleg. Alvaro Obregón,
01000 MEXICO, D.F.*

*IVAR MASSABO. Università della Calabria, Dipartimento di Matematica,
87036 Arcavacata di Rende (CS), ITALY.*

*ALFONSO VIGNOLI. Università di Roma Tor Vergata, Dipartimento di Matematica,
Via Fontanile di Carcaricola, 00133 ROMA, ITALY.*

Editorial Information

To be published in the *Memoirs*, a paper must be correct, new, nontrivial, and significant. Further, it must be well written and of interest to a substantial number of mathematicians. Piecemeal results, such as an inconclusive step toward an unproved major theorem or a minor variation on a known result, are in general not acceptable for publication. *Transactions* Editors shall solicit and encourage publication of worthy papers. Papers appearing in *Memoirs* are generally longer than those appearing in *Transactions* with which it shares an editorial committee.

As of September 1, 1992, the backlog for this journal was approximately 9 volumes. This estimate is the result of dividing the number of manuscripts for this journal in the Providence office that have not yet gone to the printer on the above date by the average number of monographs per volume over the previous twelve months. (There are 6 volumes per year, each containing about 3 or 4 numbers.)

A Copyright Transfer Agreement is required before a paper will be published in this journal. By submitting a paper to this journal, authors certify that the manuscript has not been submitted to nor is it under consideration for publication by another journal, conference proceedings, or similar publication.

Information for Authors

Memoirs are printed by photo-offset from camera copy fully prepared by the author. This means that the finished book will look exactly like the copy submitted.

The paper must contain a *descriptive title* and an *abstract* that summarizes the article in language suitable for workers in the general field (algebra, analysis, etc.). The *descriptive title* should be short, but informative; useless or vague phrases such as "some remarks about" or "concerning" should be avoided. The *abstract* should be at least one complete sentence, and at most 300 words. Included with the footnotes to the paper, there should be the 1991 *Mathematics Subject Classification* representing the primary and secondary subjects of the article. This may be followed by a list of *key words and phrases* describing the subject matter of the article and taken from it. A list of the numbers may be found in the annual index of *Mathematical Reviews*, published with the December issue starting in 1990, as well as from the electronic service e-MATH [**telnet e-MATH.ams.org** (or **telnet 130.44.1.100**). Login and password are **e-math**]. For journal abbreviations used in bibliographies, see the list of serials in the latest *Mathematical Reviews* annual index. When the manuscript is submitted, authors should supply the editor with electronic addresses if available. These will be printed after the postal address at the end of each article.

Electronically-prepared manuscripts. The AMS encourages submission of electronically-prepared manuscripts in $\mathcal{A}_{\mathcal{M}}\mathcal{S}$-TEX or $\mathcal{A}_{\mathcal{M}}\mathcal{S}$-LATEX. To this end, the Society has prepared "preprint" style files, specifically the amsppt style of $\mathcal{A}_{\mathcal{M}}\mathcal{S}$-TEX and the amsart style of $\mathcal{A}_{\mathcal{M}}\mathcal{S}$-LATEX, which will simplify the work of authors and of the production staff. Those authors who make use of these style files from the beginning of the writing process will further reduce their own effort.

Guidelines for Preparing Electronic Manuscripts provide additional assistance and are available for use with either $\mathcal{A}_{\mathcal{M}}\mathcal{S}$-TEX or $\mathcal{A}_{\mathcal{M}}\mathcal{S}$-LATEX. Authors with FTP access may obtain these *Guidelines* from the Society's Internet node e-MATH.ams.org (130.44.1.100). For those without FTP access they can be obtained free of charge from the e-mail address guide-elec@math.ams.org (Internet) or from the Publications Department, P. O. Box 6248, Providence, RI 02940-6248. When requesting *Guidelines* please specify which version you want.

Electronic manuscripts should be sent to the Providence office only after the paper has been accepted for publication. Please send electronically prepared manuscript files via e-mail to pub-submit@math.ams.org (Internet) or on diskettes to the Publications Department address listed above. When submitting electronic manuscripts please be sure to include a message indicating in which publication the paper has been accepted.

For papers not prepared electronically, model paper may be obtained free of charge from the Editorial Department at the address below.

Two copies of the paper should be sent directly to the appropriate Editor and the author should keep one copy. At that time authors should indicate if the paper has been prepared using $\mathcal{A}_{\mathcal{M}}\mathcal{S}$-TEX or $\mathcal{A}_{\mathcal{M}}\mathcal{S}$-LATEX. The *Guide for Authors of Memoirs* gives detailed information on preparing papers for *Memoirs* and may be obtained free of charge from AMS, Editorial Department, P. O. Box 6248, Providence, RI 02940-6248. The *Manual for Authors of Mathematical Papers* should be consulted for symbols and style conventions. The *Manual* may be obtained free of charge from the e-mail address cust-serv@math.ams.org or from the Customer Services Department, at the address above.

Any inquiries concerning a paper that has been accepted for publication should be sent directly to the Editorial Department, American Mathematical Society, P. O. Box 6248, Providence, RI 02940-6248.